An Introduction

to the

ELASTIC STABILITY

OF STRUCTURES

PRENTICE-HALL CIVIL ENGINEERING
AND ENGINEERING MECHANICS SERIES

N. M. Newmark and W. J. Hall,
Editors

An Introduction
to the
ELASTIC STABILITY
OF STRUCTURES

GEORGE J. SIMITSES

Professor
School of Engineering Science and Mechanics
Georgia Institute of Technology

PRENTICE-HALL, INC., Englewood Cliffs, New Jersey

Library of Congress Cataloging in Publication Data

SIMITSES, GEORGE J.
An introduction to the elastic stability of structures.

(Civil engineering and engineering mechanics)
Includes bibliographical references.
1. Structural stability. 2. Buckling (Mechanics)
3. Elasticity. I. Title.
TA656.S55   624′.176   74–30479
ISBN 0–13–481200–X

© 1976 by Prentice-Hall, Inc.
Englewood Cliffs, New Jersey

10   9   8   7   6   5   4   3   2   1

Printed in the United States of America

PRENTICE-HALL INTERNATIONAL, INC., *London*
PRENTICE-HALL OF AUSTRALIA, PTY. LTD., *Sydney*
PRENTICE-HALL OF CANADA, LTD., *Toronto*
PRENTICE-HALL OF INDIA PRIVATE LIMITED, *New Delhi*
PRENTICE-HALL OF JAPAN, INC., *Tokyo*

To my children John, William, and Alexandra,
my wife Nena,
and my parents John and Vasilike

CONTENTS

PREFACE

Knowledge of structural stability theory is of paramount importance to the practicing structural engineer. In many instances, buckling is the primary consideration in the design of various structural configurations. Because of this, formal courses in this important branch of mechanics are available to students in Aerospace Engineering, Civil Engineering, Engineering Science and Mechanics, and Mechanical Engineering at many institutions of higher learning. This book is intended to serve as a text in such courses. The emphasis of the book is on the fundamental concepts and on the methodology developed through the years to solve structural stability problems.

The material contained in this text is ideally suited for a one-semester Master's level course, although with judicious addition or deletion of topics, the text may be adopted for both a two-quarter series or a one-quarter course.

The first chapter introduces the basic concepts of elastic stability and the approaches used in solving stability problems, it also discusses the different buckling phenomena that have been observed in nature. In Chapter 2, the basic concepts and methodology are applied to some simple mechanical models with finite degrees of freedom. This is done to help the student understand the fundamentals without getting involved with lengthy and complicated mathematical operations, which is usually the case when dealing with the continuum (infinitely many degrees of freedom). In Chapter 3, a complete treatment of the elastic stability of columns is presented, including effects of elastic restraints. Some simple frame problems are discussed in

Chapter 4. This chapter is of special importance to the Civil Engineering student. Since energy-based methods have been successfully used in structural mechanics, Chapter 5 presents a comprehensive treatment of the energy criterion for stability and contains many energy-related methods. The study of this chapter requires some knowledge of work- and energy-related principles and theorems. These topics are presented in the Appendix for the benefit of the student who never had a formal course in this area. Columns on elastic foundations are discussed in Chapter 6. Chapter 7 presents a comprehensive treatment of the buckling of thin rings and high and low arches. In this chapter, a complete analysis is given for a shallow, pinned, sinusoidal arch on an elastic foundation subject to a sinusoidal transverse loading. This is an interesting model for stability studies because, depending upon the values of the different parameters involved, it exhibits all types of buckling that have been observed in different structural systems: top-of-the-knee buckling, stable bifurcation (Euler-type), and unstable bifurcation. Finally, Chapter 8 contains some remarks about stability of nonconservative elastic systems and dynamic stability. The purpose of this chapter is to motivate the student for further studies by using the references cited.

Once the student has been exposed to the contents of this text, he may, depending upon his interest, proceed with the study of the stability analysis of other structural configurations such as plates, shells, and torsional and lateral buckling of thin-walled open-section beams.

Numerous references are listed at the end of each chapter. These references provide an excellent source for further studies, for better understanding of certain specific concepts, and for detailed information about specific applications.

The author is indebted to the late Professor J. N. Goodier whose two-course series at Stanford University provided the basis for the organization of the material in the present text. The encouragement and valuable suggestions of Professor N. J. Hoff are greatly appreciated. Special thanks are due to Professor M. E. Raville for providing tangible and intangible support, reading the manuscript, and making many corrections. The numerous discussions with Professors S. Atluri, W. W. King, G. M. Rentzepis, C. V. Smith, Jr., and M. Stallybrass are gratefully acknowledged. The proofreading was done by many of my students, but special thanks are due particularly to Dr. V. Ungbhakorn and Mr. J. Giri. Mr. Giri also made most of the drawings. Mrs. Ruth Salley and Mrs. Jackie Van Hook worked with great dedication in typing the manuscript.

G.J.S.

An Introduction
to the
ELASTIC STABILITY
OF STRUCTURES

1

INTRODUCTION

AND FUNDAMENTALS

1.1 MOTIVATION

Many problems are associated with the design of modern structural systems. Economic factors, availability and properties of materials, interaction between the external loads (aerodynamic) and the response of the structure, dynamic and temperature effects, performance, cost, and ease of maintenance of the system are all problems which are closely associated with the synthesis of these large and complicated structures. Synthesis is the branch of engineering which deals with the design of a system for a given mission. Synthesis requires the most efficient manner of designing a system (i.e., most economical, most reliable, lightest, best, and most easily maintained system), and this leads to *optimization*. An important part of system optimization is structural optimization, which is based on the assumption that certain parameters affecting the system optimization are given (i.e., overall size and shape, performance, nonstructural weight, etc.) It can only be achieved through good theoretical analyses supported by well-planned and well-executed experimental investigations.

Structural analysis is that branch of structural mechanics which associates the behavior of a structure or structural elements with the action of external causes. Two important questions are usually asked in analyzing a structure: (1) What is the response of the structure when subjected to external causes (loads and temperature changes)? In other words, if the external causes are known, can we find the deformation patterns and the internal load distribution? (2) What is the character of the response? Here we are interested in knowing if the equilibrium is stable or if the motion is limited (in the case of dynamic causes). For example, if a load is periodically applied, will the structure oscillate within certain bounds or will it tend to move without bounds?

If the dynamic effects are negligibly small, in which case the loads are said to be applied quasistatically, then the study falls in the domain of structural *statics*. On the other hand, if the dynamic effects are not negligible, we are dealing with structural *dynamics*.

The branch of structural statics that deals with the character of the response is called stability or instability of structures. The interest here lies in the fact that stability criteria are often associated directly with the load-carrying capability of the structure. For example, in some cases instability is not directly associated with the failure of the overall system, i.e., if the skin wrinkles, this does not mean that the entire fuselage or wing will fail. In other cases though, if the portion of the fuselage between two adjacent rings becomes unstable, the entire fuselage will fail catastrophically. Thus, stability of structures or structural elements is an important phase of structural analysis, and consequently it affects structural synthesis and optimization.

1.2 STABILITY OR INSTABILITY OF STRUCTURES

There are many ways a structure or a structural element can become unstable, depending on the structural geometry and the load characteristics. The spatial geometry, the material along with its distribution and properties, the character of the connections (riveted joints, welded, etc.), and the supports comprise the structural geometry. By load characteristics we mean spatial distribution of the load, load behavior (whether or not the load is affected by the deformation of the structure, e.g., if a ring is subjected to uniform radial pressure, does the load remain parallel to its initial direction, does it remain normal to the deformed ring, or does it remain directed towards the initial center of curvature?), and/or whether the force system is conservative.

1.2-1 Conservative Force Field

A mechanical system is conservative if subjected to conservative forces. If the mechanical system is rigid, there are only external forces; if the system is deformable, the forces may be both external and internal. Regardless of the composition, a system is conservative if all the forces are conservative. A force acting on a mass particle is said to be conservative if the work done by the force in displacing the particle from position 1 to position 2 is independent of the path. In such a case, the force may be derived from a potential. A rigorous mathematical treatment is given below for the interested student.

The work done by a force $\vec{F}$ acting on a mass particle in moving the particle from position P_0 (at time t_0) to position P_1 (at time t_1) is given by

$$W = \underset{C}{\oint}_{\vec{r_0}}^{\vec{r_1}} \vec{F} \cdot d\vec{r} \tag{1}$$

Thus the integral, W (a scalar), depends on the initial position, $\vec{r_0}$, the final position, $\vec{r_1}$, and the path C. If a knowledge of the path C is not needed and the work is a function of the initial and final positions only, then

$$W = W(\vec{r_0}, \vec{r_1}, \vec{F}) \tag{2}$$

and the force field is called *conservative*. (See Refs. 1–3.)

Parenthesis. If S denotes some surface in the space and C some space curve, then by *Stokes' theorem*

$$\oint_C \vec{U} \cdot d\vec{l} = \iint_S \operatorname{curl} \vec{U} \cdot \vec{n} \, ds \tag{3}$$

where $\vec{n}$ is a unit vector normal to the surface S (see Fig. 1-1).

If $\oint_C \vec{U} \cdot d\vec{l} = 0$, then

$$\iint_S \operatorname{curl} \vec{U} \cdot \vec{n} \, ds = 0 \tag{4}$$

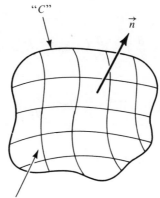

Figure 1-1. "S"

for all surfaces S and spanning curves C. If this is so, then the curl of $\vec{U}$ (some vector quantity) must be identically equal to zero, or

$$\text{curl } \vec{U} \equiv 0 \tag{5}$$

Next, if we apply this result to a conservative force field where $\vec{U}$ is replaced by $\vec{F}$, then according to the previous result

$$\text{curl } \vec{F} \equiv 0$$

It is well known from *vector analysis* that the curl of the gradient of any scalar function vanishes identically. Therefore, for a conservative field we may write

$$\vec{F} = -\nabla V \tag{6}$$

where:

1. The negative sign is arbitrary,
2. V is some scalar function, and
3. ∇ is the vector operator

$$\frac{\partial}{\partial x}\vec{i} + \frac{\partial}{\partial y}\vec{j} + \frac{\partial}{\partial z}\vec{k}$$

where $\vec{i}, \vec{j}, \vec{k}$ form an orthogonal unit vector triad along x, y, z, respectively.

This implies that the force can be derived from a potential.

Note that in this case the work done by the force in a conservative force field is given by

$$W = \oint_{r_0}^{\vec{r}_1} \vec{F} \cdot d\vec{r} = -\oint_{r_0}^{\vec{r}_1} \nabla V \cdot d\vec{r} = -\oint_{r_0}^{\vec{r}_1} (\nabla \cdot d\vec{r}) V$$

and since

$$\nabla = \frac{\partial}{\partial x}\vec{i} + \frac{\partial}{\partial y}\vec{j} + \frac{\partial}{\partial z}\vec{k} \quad \text{and} \quad d\vec{r} = (dx)\vec{i} + (dy)\vec{j} + (dz)\vec{k}$$

then

$$(\nabla \cdot d\vec{r}) V = \frac{\partial V}{\partial x} dx + \frac{\partial V}{\partial y} dy + \frac{\partial V}{\partial z} dz = dV$$

or

$$W = -\int_{V_0}^{V_1} dV = V_0 - V_1 = -\delta(V) \tag{7}$$

where δ denotes a change in the potential of the conservative force $\vec{F}$ from position $\vec{r}_0$ to position $\vec{r}_1$.

Thus a system is conservative if the work done by the forces in displacing the system from deformation state 1 to deformation state 2 is independent of the path. If this is the case, the force can be derived from a potential.

There are many instances where systems are subjected to loads which cannot be derived from a potential. For instance, consider a column clamped at one end and subjected to an axial load at the other, the direction of which is tangential to the free end at all times (follower force). Such a system is nonconservative and can easily be deduced if we consider two or more possible paths that the load can follow in order to reach a final position. In each case the work done will be different. Systems subject to time-dependent loads are also nonconservative. Nonconservative systems have been given special consideration (Refs. 4 and 5), and the emphasis in this text will be placed on conservative systems (see Ref. 6 for a detailed description of forces and systems).

1.2-2 The Concept of Stability

As the external causes are applied quasistatically, the elastic structure deforms and static equilibrium is maintained. If now at any level of the external causes "small" external disturbances are applied and the structure reacts by simply performing oscillations about the deformed equilibrium state, the equilibrium is said to be *stable*. The disturbances can be in the form of deformations or velocities, and by "small" we mean as small as desired. As a result of this latter definition, it would be more appropriate to say that the equilibrium is stable in the small. In addition, when the disturbances are applied, the level of the external causes is kept constant. On the other hand, if the elastic structure either tends to and does remain in the disturbed position or tends to and/or diverges from the deformed equilibrium state, the equilibrium is said to be *unstable*. Some authors prefer to distinguish these two conditions and call the equilibrium *neutral* for the former case and *unstable* for the latter. When either of these two cases occurs, the level of the external causes is called *critical*.

This can best be demonstrated by the system shown in Fig. 1-2. This system consists of a ball of weight W resting at different points on a surface with zero curvature normal to the plane of the figure. Points of zero slope on the

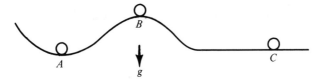

Figure 1-2. Character of static equilibrium positions.

surface denote positions of static equilibrium (points *A*, *B*, and *C*). Furthermore, the character of equilibrium at these points is substantially different. At *A*, if the system is disturbed through infinitesimal disturbances (small displacements or small velocities), it will simply oscillate about the static equilibrium position *A*. Such equilibrium position is called *stable* in the small. At point *B*, if the system is disturbed, it will tend to move away from the static equilibrium position *B*. Such an equilibrium position is called *unstable* in the small. Finally, at point *C*, if the system is disturbed, it will tend to remain in the disturbed position. Such an equilibrium position is called *neutrally stable* or *indifferent* in the small. The expression "in the small" is used because the definition depends on the *small* size of the perturbations. If the disturbances are allowed to be of finite magnitude, then it is possible for a system to be unstable in the small but stable in the large (point *B*, Fig. 1-3a) or stable in the small but unstable in the large (point *A*, Fig. 1-3b).

In most structures or structural elements, loss of stability is associated with the tendency of the configuration to pass from one deformation pattern to another. For instance, a long, slender column loaded axially, at the critical condition, passes from the straight configurations (pure compression) to the combined compression and bending state. Similarly, a perfect, complete, thin, spherical shell under external hydrostatic pressure, at the critical condition, passes from a pure membrane state (uniform radial displacement only; shell stretching) to a combined stretching and bending state (nonuniform radial displacements). This characteristic has been recognized for many years and it was first used to solve stability problems of elastic structures. It allows the analyst to reduce the problem to an eigenvalue problem, and many names have been given to this approach: the classical method, the bifurcation method, the equilibrium method, and the static method.

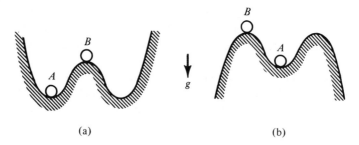

(a)   (b)

Figure 1-3. Character of static equilibrium positions in the large.

1.2-3 Critical Loads Versus Buckling Load

At this point nomenclature merits some attention. There is a definite difference in principle between the buckling load observed in a loading process where the loads keep changing (observed physical phenomenon) and the buckling load calculated from some mathematical model, which always refers to a system with prescribed loads. Since the latter is based on theory and is usually obtained as the characteristic or eigenvalue of some eigen-boundary-value problem, it is properly called the critical load.

In the process of buckling in the testing machine, in the static or dynamic testing of a structural configuration, and in the failure of the structure in actual use, we are confronted with the physical aspects of buckling. The load at which a structure buckles should preferably be designated as the buckling load.

The compound term *critical buckling load* is unnecessary and should be avoided. It may have originated from the observation that theory (for the ideal column, for instance) predicts several critical loads (eigenvalues) corresponding to different deflection patterns (eigenfunctions). In an experiment, however, only one buckling pattern is observed, namely, the one that corresponds to the lowest eigenvalue. This lowest eigenvalue is no more critical than any of the higher ones, but it is the one that corresponds to the observed buckling load. If it is desired to give it a special designation, it should be called the lowest critical load, rather than the critical buckling load.

1.2-4 Basic Approaches of Stability Analysis

A number of approaches have been successfully used in determining critical conditions for elastic structures which are subject to instability. The oldest approach, which is applicable to many problems, is concerned with the answer to the following question. If an external cause is applied quasi-statically to an elastic structure, is there a level of the external cause at which two or more different but infinitesimally close equilibrium states can exist? By different equilibrium states we mean that the response of the structure is such that equilibrium can be maintained with different deformation patterns. An example of this is the long perfect column loaded axially in compression. As the load increases quasistatically from zero, the column is compressed but remains straight. At some value of the load though, a bent position of infinitesimal amplitude also represents an equilibrium position. Since at this value of the load there are two different equilibrium states infinitesimally close, a bifurcation point exists (adjacent equilibrium positions). Mathematically, in this approach, the problem is reduced to an eigen-boundary-value problem and the critical conditions are denoted by the eigenvalues. This approach is usually referred to as the *classical* approach, *equilibrium* approach, or *bifurcation* approach. Many examples will be discussed in the chapters to follow.

Another approach is to write the equations governing small free vibrations of the elastic structures at some level of the external causes (treated as a constant) and try to find out for what level of the external cause the motion ceases to be bounded in the small. In writing the governing equations, one must allow all possible modes of deformation. The form of equilibrium is said to be stable if a slight disturbance (in the form of displacement or velocity) causes a small deviation of the system from the considered equilibrium configuration, but by decreasing the magnitude of the disturbance, the deviation can be made as small as required. On the other hand, a critical condition is reached if a disturbance, however small, causes a finite deviation of the system from the considered form of equilibrium. This approach is known as the *kinetic* or *dynamic* approach, and it is a direct application of the stability concept demonstrated in Fig. 1-2.

Next, if a system is conservative, the forces can be derived from a potential, and the total potential of the system can be expressed in terms of the generalized coordinates and the external forces. The generalized coordinates are the parameters needed to express the deflectional shapes which the elastic structure could possibly assume. In this case, the equilibrium is stable in the small if the total potential is a relative minimum. This approach is completely equivalent to the kinetic approach (a proof is given in Ref.1) for conservative systems, and it is known as the *potential energy* method or simply the *energy* method. This definition of stability requires special attention and it will be fully justified in the next section.

Finally, there is a fourth approach in dealing with stability problems of elastic structures. This method is usually called the *imperfection* method. The question in this case is: "What is the value of the load (level of external causes) for which the deflections of an imperfect system increase beyond any limit?" It should be pointed out that certain systems, when subjected to certain external causes, are imperfection sensitive. This means that the critical conditions of the perfect system are different from those of the imperfect one. Imperfection sensitivity has served to explain the discrepancy between theory and experiment for such systems. It will also be demonstrated that there are systems for which the perfect and imperfect systems have the same critical conditions according to the approaches defined above. It is the opinion of the author that the imperfection approach should not be associated with the stability of the perfect system, but simply characterize the response of the imperfect system. In short, the stability of a system, whether perfect or imperfect, should be investigated by the first three methods (whichever is applicable).

1.2-5 The Energy Method

This method is based on the kinetic criterion of stability and it is an association of this criterion with characteristics of the total potential (relative minimum) surface at a position of static equilibrium. Since it requires the existence of a total potential surface, this method is applicable only to conservative systems.

Before the energy criterion is justified, let us describe in analytical form the kinetic criterion of stability. This concept was first introduced by Lagrange (Ref. 7) for a system with a finite number of degrees of freedom.

A more strict definition of stability of equilibrium was given by Lyapunov (see Refs. 3, 8, 9, 10, and 11) as a particular case of motion. Let us assume that the position of a system depends on n generalized coordinates $q_i (i = 1, 2, \ldots, n)$ and that a static equilibrium state is characterized by $q_i = 0$. Let the system be at this static equilibrium position, and at time $t = 0$ we allow small bounded disturbances $|q_i^0| < \delta$ and $|\dot{q}_i^0| < \delta$. The response of the system at any instant $t > 0$ is characterized by $q_i(t)$ and $\dot{q}_i(t)$. If the response is also bounded

$$|q_i(t)| < \epsilon \quad \text{and} \quad |\dot{q}_i(t)| < \epsilon \tag{8}$$

then we say that the static equilibrium position $q_i = 0$ is stable. In other words, in the case of stable static equilibrium (in the small) positions, we can always select such small initial conditions that the generalized coordinates and velocities are bounded.

The energy criterion is based on the Lagrange-Dirichlet theorem, which states: If the total potential has a relative minimum at an equilibrium position (stationary value), then the equilibrium position is stable. This theorem can easily be proven if we simply employ the principle of conservation of energy for conservative systems, which states that the sum of the kinetic energy and the total potential is a constant $(T + U = c)$. Now if we define the equilibrium position by $q_i = 0$ and let $U(0) = 0$, then, if $U(0)$ is a minimum, $U(q_i)$ must have a positive lower bound $\bar{c}$ on the boundary of any sufficiently close neighborhood of $q_i = 0$. It is now always possible to select q_i^0 and $\dot{q}_i^0$ such that $T + U = c$ and $c < \bar{c}$. In other words, since the sum of the total potential and the nonnegative kinetic energy is a constant c, if $c < \bar{c}$ the boundary of the neighborhood of $q_i = 0$ can never be reached and the equilibrium position $q_i = 0$ is stable (bounded motion). Unfortunately, it is very difficult to prove the converse of the Lagrange-Dirichlet theorem. A statement of this converse theorem is as follows: If the equilibirum is stable at an equilibrium position characterized by $q_i = 0$, then $U(0)$ is a relative minimum. Proof of this theorem under certain restrictive assumptions has been given by Chetayev (Ref. 12). Although there is no general proof of this converse theorem, its validity has been accepted and the energy criterion has been used as both a necessary and sufficient condition for stability. This criterion for stability can be generalized for systems with infinitely many degrees of freedom (cohesive, continuous, deformable configurations).

The energy criterion can be used to arrive at critical conditions by simply seeking load conditions at which the response of the system ceases to be in stable equilibrium. This implies that we are interested in knowing explicitly the conditions under which the change in the total potential is positive definite. If the total potential is expressed as a Taylor series about the static equilibrium point characterized by $q_i = 0$, then

$$U(q_1, q_2, \ldots, q_N) = U(0, 0, \ldots, 0) + \sum_{i=1}^{N} \frac{\partial U}{\partial q_i}\bigg|_0 q_i$$
$$+ \frac{1}{2} \sum_{i=1}^{N} \sum_{j=1}^{N} \frac{\partial^2 U}{\partial q_i\, \partial q_j}\bigg|_0 q_i q_j + \cdots \tag{9}$$

Since $q_i = 0$ characterizes a position of static equilibrium, then

$$\frac{\partial U}{\partial q_i}\bigg|_0 = 0 \tag{10}$$

and

$$U(q_1, q_2, \ldots, q_N) - U(0, 0, \ldots, 0) = \Delta U = \tfrac{1}{2} \sum_{i=1}^{N} \sum_{j=1}^{N} c_{ij} q_i q_j \tag{11}$$

where

$$c_{ij} = \frac{\partial^2 U}{\partial q_i\, \partial q_j}\bigg|_0$$

The energy criterion requires that the homogeneous quadratic form given by Eq. (11) be positive definite.

THEOREM   The homogeneous quadratic form

$$U(q_1, q_2, \ldots, q_N) = \tfrac{1}{2} \sum_{i=1}^{N} \sum_{j=1}^{N} c_{ij} q_i q_j \tag{12}$$

is positive definite if and only if the determinant D of its coefficients, c_{ij}, and its principal minors, D_i, are all positive.

$$
\begin{array}{c}
D_1 \quad D_2 \quad D_3 \;\cdots \\[4pt]
\begin{vmatrix}
c_{11} & c_{12} & c_{13} & c_{14} & \cdots & c_{1N} \\
c_{21} & c_{22} & c_{23} & c_{24} & \cdots & c_{2N} \\
c_{31} & c_{32} & c_{33} & c_{34} & \cdots & c_{3N} \\
\vdots & \vdots & \vdots & \vdots & & \vdots \\
c_{N1} & c_{N2} & c_{N3} & c_{N4} & \cdots & c_{NN}
\end{vmatrix} > 0
\end{array}
\tag{13}
$$

Proof: The proof will be given in a number of steps.

1. If U is positive for any set of coordinates $[q_i] \neq [0]$ (not all zero), then

$$U(q_1, 0, 0, 0, \ldots, 0) = \tfrac{1}{2}c_{11}q_1^2 > 0 \tag{14}$$

which requires that $c_{11} > 0$. Note that if c_{11} is positive, then $U(q_1, 0, 0, \ldots, 0) > 0$.

2. Assuming that $c_{11} \neq 0$, we can make the following transformation:

$$q_1^* = q_1 + \frac{c_{12}}{c_{11}}q_2 + \frac{c_{13}}{c_{11}}q_3 + \cdots + \frac{c_{1N}}{c_{11}}q_N$$

$$= q_1 + \sum_{i=2}^{N} \frac{c_{1i}}{c_{11}}q_i \tag{15}$$

With this transformation we note that

$$\frac{1}{2}c_{11}(q_1^*)^2 = \frac{1}{2}c_{11}\left(q_1 + \sum_{i=2}^{N} \frac{c_{1i}}{c_{11}}q_i\right)^2$$

$$= \frac{1}{2}c_{11}q_1^2 + q_1 \sum_{i=2}^{N} c_{1i}q_i + \frac{1}{2}\sum_{i=2}^{N}\sum_{j=2}^{N} \frac{c_{1i}c_{1j}}{c_{11}}q_iq_j \tag{16}$$

From Eq. (16)

$$\frac{1}{2}c_{11}q_1^2 = \frac{1}{2}c_{11}(q_1^*)^2 - \left[q_1\sum_{i=2}^{N}c_{1i}q_i + \frac{1}{2}\sum_{i=2}^{N}\sum_{j=2}^{N}\frac{c_{1i}c_{1j}}{c_{11}}q_iq_j\right] \tag{17}$$

Next we rewrite Eq. (12) in the following form:

$$U(q_1, q_2, \ldots, q_N) = \tfrac{1}{2}c_{11}q_1^2 + q_1 \sum_{i=2}^{N} c_{1i}q_i + \tfrac{1}{2}\sum_{i=2}^{N}\sum_{j=2}^{N} c_{ij}q_iq_j \tag{18}$$

Substitution of Eq. (17) into Eq. (18) yields

$$U(q_1^*, q_2, \ldots, q_N) = \frac{1}{2}c_{11}q_1^{*2} + \frac{1}{2}\sum_{i=2}^{N}\sum_{j=2}^{N}\left[c_{ij} - \frac{c_{1i}c_{1j}}{c_{11}}\right]q_iq_j \tag{19}$$

If we let

$$c_{ij} - \frac{c_{1i}c_{1j}}{c_{11}} = \alpha_{ij} \tag{20}$$

then Eq. (19) becomes

$$U(q_1^*, q_2, \ldots, q_N) = \tfrac{1}{2}c_{11}q_1^{*2} + \tfrac{1}{2}\sum_{i=2}^{N}\sum_{j=2}^{N}\alpha_{ij}q_iq_j \tag{21}$$

3. If U is positive for $q_1^* \neq 0$ and $q_i = 0$ ($i = 2, 3, \ldots, N$), then $c_{11} > 0$. If U is positive for $q_1^* = 0$, $q_2 \neq 0$, and $q_i = 0$ ($i = 3, \ldots, N$), then $\alpha_{22} > 0$.

Note that the converse is also true for the same condition, i.e., if c_{11} is positive, U is positive, and if α_{22} is positive, U is positive.

These conditions for positive U can be written solely in terms of c_{ij} by use of Eq. (20), or

$$c_{11} > 0 \quad \text{and} \quad c_{11}c_{22} - c_{12}^2 > 0 \tag{22}$$

Note that the second inequality is equivalent to the requirement $D_2 > 0$ if $c_{12} = c_{21}$. This requirement is by no means restrictive since Eq. (12) represents a homogeneous quadratic form.

4. Next step 2 is repeated with $c_{22} \neq 0$ and the following transformation:

$$q_2^* = q_2 + \sum_{i=3}^{N} \frac{\alpha_{2i}}{\alpha_{22}} q_i \tag{23}$$

This transformation leads to the following expression for U:

$$U(q_1^*, q_2^*, q_3, \ldots, q_N) = \tfrac{1}{2}c_{11}q_1^{*2} + \tfrac{1}{2}\alpha_{22}q_2^{*2} + \tfrac{1}{2}\sum_{i=3}^{N}\sum_{j=3}^{N} \beta_{ij}\, q_i q_j \tag{24}$$

where

$$\beta_{ij} = \alpha_{ij} - \frac{\alpha_{2i}\alpha_{2j}}{\alpha_{22}} \tag{25}$$

As in step 3

$$U(q_1^*, 0, 0, \ldots, 0) > 0 \quad \text{if and only if} \quad c_{11} > 0$$
$$U(0, q_2^*, 0, 0, \ldots, 0) > 0 \quad \text{if and only if} \quad \alpha_{22} > 0$$

and

$$U(0, 0, q_3, 0, \ldots, 0) > 0 \quad \text{if and only if} \quad \beta_{33} > 0$$

This requirement implies that

$$\alpha_{22}\alpha_{33} - \alpha_{23}^2 > 0 \tag{26}$$

By Eq. (20)

$$\left(c_{22} - \frac{c_{12}^2}{c_{11}}\right)\left(c_{33} - \frac{c_{13}^2}{c_{11}}\right) - \left(c_{23} - \frac{c_{12}c_{13}}{c_{11}}\right)^2 > 0 \tag{27}$$

This last requirement is equivalent to $D_3 > 0$ provided $c_{ij} = c_{ji}$.

5. The continuation of this procedure eventually leads to the representation of the homogeneous quadratic form as a linear combination of squares:

$$U = \tfrac{1}{2}c_{11}q_1^{*2} + \tfrac{1}{2}\alpha_{22}q_2^{*2} + \tfrac{1}{2}\beta_{33}q_3^{*2} + \cdots \tag{28}$$

From this form, it is clearly seen that U is positive definite if and only if

$$c_{11} > 0, \qquad \alpha_{22} > 0, \qquad \beta_{33} > 0 \qquad \text{QED} \tag{29}$$

Use of this theorem in the energy criterion implies that a position of static equilibrium is stable if and only if

$$D_n = \begin{vmatrix} \dfrac{\partial^2 U_T}{\partial q_1^2} & \dfrac{\partial^2 U_T}{\partial q_1\,\partial q_2} & \dfrac{\partial^2 U_T}{\partial q_1\,\partial q_3} & \cdots & \dfrac{\partial^2 U_T}{\partial q_1\,\partial q_N} \\[2mm] \dfrac{\partial^2 U_T}{\partial q_2\,\partial q_1} & \dfrac{\partial^2 U_T}{\partial q_2^2} & \dfrac{\partial^2 U_T}{\partial q_2\,\partial q_3} & \cdots & \dfrac{\partial^2 U_T}{\partial q_2\,\partial q_N} \\[2mm] \dfrac{q^2 U_T}{\partial q_3\,\partial q_1} & \dfrac{\partial^2 U_T}{\partial q_3\,\partial q_2} & \dfrac{\partial^2 U_T}{\partial q_3^2} & \cdots & \dfrac{\partial^2 U_T}{\partial q_3\,\partial q_N} \\[2mm] \cdot & \cdot & \cdot & & \cdot \\ \cdot & \cdot & \cdot & & \cdot \\ \cdot & \cdot & \cdot & & \cdot \\[2mm] \dfrac{\partial^2 U_T}{\partial q_N\,\partial q_1} & \dfrac{\partial^2 U_T}{\partial q_N\,\partial q_2} & \dfrac{\partial^2 U_T}{\partial q_N\,\partial q_3} & \cdots & \dfrac{\partial^2 U_T}{\partial q_N^2} \end{vmatrix} > 0 \tag{30}$$

with column headings D_1, D_2, D_3, D_N

and all its principal minors $D_1 > 0$, $D_2 > 0$, etc.

In all problems in mechanics, dealing with the stability of elastic systems under external causes, the total potential of the system depends not only on the generalized coordinates (variables defining the position of the system) but also on certain parameters that characterize the external cause or causes.

The general theory of equilibrium positions of such systems with various values of the parameters was established by Poincaré (Ref. 13; see also Ref. 12). Among the findings of Poincaré are the following (simplified in this text for the sake of understanding):

1. The requirements

$$\frac{\partial U_T}{\partial q_i} = 0 \quad \text{and} \quad D_n = 0$$

define a point of bifurcation (intersection of static equilibrium branches at the same value of the external cause parameter). See for example Figs. 1-4 and 1-5 (points A and A').

2. Changes in stability along the primary path (from stable to unstable equilibrium positions) do occur at points of bifurcation. Consider, for example, branch OAB of Fig. 1-4. If the part of this branch characterized by OA denotes stable static equilibrium positions, the part characterized by AB must denote unstable static equilibrium positions.

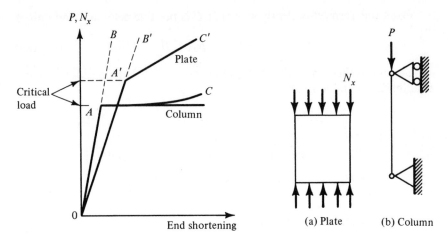

Figure 1-4. Classical buckling.

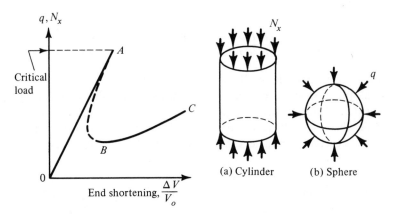

Figure 1-5. Finite-disturbance buckling.

These findings support the classical approach to stability problems which only seeks bifurcation points. The external cause condition at such a point is called a *critical condition*.

1.2-6 Types of Buckling

When the external causes are applied quasistatically and the level at which instability occurs is reached, the elastic structure assumes an equilibrium configuration which is distinctly different from the ones assumed during the quasistatic application of the causes. When this occurs, we say that the elastic structure has *buckled*. Since there are different ways by which the new equilibrium configuration may be reached, buckling can be classified by the use of proper adjectives.

The type of buckling that was first studied and has been given the most attention is the so-called *classical or bifurcation buckling.* This type of buckling is characterized by the fact that, as the load passes through its critical stage, the structure passes from its unbuckled equilibrium configuration to an infinitesimally close buckled equilibrium configuration. As will be demonstrated in later chapters, buckling of long straight columns loaded axially, buckling of thin plates loaded by in-plane loads, and buckling of rings are classical examples of this kind of buckling (see Fig. 1-4).

Another type of buckling is what Libove (Ref. 14) calls *finite-disturbance buckling.* For some structures, the loss of stiffness after buckling is so great that the buckled equilibrium configuration can only be maintained by returning to an earlier level of loading. Classical examples of this type are buckling of thin cylindrical shells under axial compression and buckling of complete, spherical, thin shells under uniform external pressure (see Fig. 1-5). In Fig. 1-5a, N_x denotes the applied axial load per unit length. In Fig. 1-5b, q denotes the uniform external pressure, V_0 the initial volume of the sphere, and ΔV the change in the volume during loading. The reason for the name is that in such structures a finite disturbance during the quasistatic application of the load can force the structure to pass from an unbuckled equilibrium configuration to a nonadjacent buckled equilibrium configuration before the classical buckling load, P_{cr}, is reached. A third type of buckling is known as *snapthrough* buckling or oil-canning (Durchschlag). This phenomenon is characterized by a visible and sudden jump from one equilibrium configuration to another equilibrium configuration for which displacements are larger than in the first (nonadjacent equilibrium states). Classical examples of this type are snapping of a low pinned arch under lateral loads (see Fig. 1-6) and snapping of clamped shallow spherical caps under uniform lateral pressure.

The above discussion shows that there is some similarity between finite-disturbance buckling and snapthrough buckling. It should also be men-

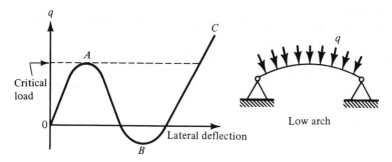

Figure 1-6. Snapthrough buckling.

tioned that, for many systems, nonlinear theory must be used to either evaluate critical conditions and/or explain the buckling phenomena.

It will become evident in subsequent chapters that there are two different viewpoints as far as types of buckling are concerned and two classifications within each viewpoint. The first viewpoint is based on the existence of a bifurcation point. For the examples shown in Figs. 1-4 and 1-5, there is a bifurcation point (A or A'). For the example shown in Fig. 1-6, there is no bifurcation point at A.

The second viewpoint is based on the expected response of the system under deadweight loading. For the examples in Fig. 1-4, the branches AC and $A'C'$ correspond to stable static equilibrium positions, and under deadweight loading there exists the possibility for the system to pass from one deformation configuration (the straight for the column) to another deformation configuration (the bent or buckled) with no appreciable dynamic effects (time-independent response). For the example shown in Fig. 1-5, since the branch AB is unstable when the system reaches point A (under deadweight loading), it will tend to snap through toward a far stable equilibrium position with a time-dependent response. This is very much the same situation for the system of Fig. 1-6. When point A is reached, the system will snap through toward a far stable equilibrium position.

In a later chapter, a model is considered which exhibits all types of buckling, top-of-the-knee (Fig. 1-6), stable bifurcation (Fig. 1-4), and unstable bifurcation (Fig. 1-5). This model is a low half-sine arch simply supported at both ends under quasistatic application of a half-sine transverse loading resting on an elastic foundation.

In investigating stability problems, one should always consider the effect of load behavior. In the case of a circular ring loaded uniformly by a radial pressure, different critical conditions are obtained depending on the behavior of the applied load. If the load behaves as hydrostatic pressure does (remains normal to the deflected shape), the critical condition is different from the case for which the load remains directed toward the center of the ring at all times (point sink). On the other hand, the effect of load behavior for certain structures is negligible.

For some systems there are certain constraints on the loading mechanism. This case can also come under the classifications of load behavior problems. For instance, suppose that the axial load on a long, straight, elastic bar is applied through a rigid bar. At buckling, the loading member (rigid bar) may tilt, and then the load behavior is such that it aggravates the situation. Such problems are known as tilt-buckling problems and they will be discussed in later chapters also.

So far, the different types of buckling of conservative systems under quasistatic application of the external causes have been discussed. We will also discuss buckling of a limited number nonconservative systems (such as the "follower-force" problem) under quasistatic application of the load. Finally, buckling can occur under dynamic application of the external causes. In general, if the load depends explicitly on time, the system is nonconservative. A large class of such problems, where the load varies sinusoidally with time, have been discussed by Bolotin (Ref. 15). This type of dynamic loading is called by Bolotin *parametric*, and the associated phenomenon of loss of stability, *parametric excitation*. Bolotin has shown that all systems which are subject to loss of stability under quasistatically applied loads are also subject to loss of stability under parametric loads.

If the dynamic load does not depend explicitly on time, the system can be conservative (see Ref. 6). Typical examples of such loads are (1) loads suddenly applied with constant magnitude and infinite duration, and (2) ideal impulsive loads. When such loads are applied to elastic structures we may ask: "Is buckling possible under such loads, and if so, what are the critical conditions?" We may note that such loads are obvious idealizations of two extreme cases of blast loading: blasts of low decay rates and large decay times and blasts of large decay rates and short decay times, respectively.

1.3 CONTINUOUS DEFORMABLE ELASTIC BODIES

A continuous body is called *deformable* if the relative distance between any two material points changes when the system is experiencing changes in the externally applied causes. The changes in the deformations and their gradients are related to the changes in the load intensities and their rates through the constitutive relations.

If the loading path, characterized by the constitutive relations, is the same as the unloading path, the continuous deformable body is called *elastic*. If, in addition, these paths are characterized by linear relations in the absence of dynamic effects (generalized Hooke's law equations), the continuous deformable body is called *linearly elastic*. Furthermore, if the properties (modulus of elasticity, Poisson's ratio, etc.) of such a body do not depend on the position of the material point, the body is termed *homogeneous*. If these properties at a material point are independent of direction, the body is called *isotropic*. If this is not so, the body is called *anisotropic*. A particular case of anisotropic elastic bodies is orthotropic elastic bodies. An elastic body is called *orthotropic* if some or all of the properties of the elastic body differ in mutually orthogonal directions.

The branch of mechanics that deals with the behavior of elastic bodies is called theory of elasticity. Since only a relatively small number of problems can be solved by means of the exact field equations of the theory of elasticity, the structural engineer is forced to make a number of simplifying assumptions in dealing with structural problems. These simplifying assumptions depend heavily on the relative dimensions of the structural element in three-dimensional space. Depending on these assumptions, all of the structural elements fall in one of the following four categories:

1. All three dimensions are of the same magnitude (spheres, short or moderate length cylinders, etc.).

2. One of the dimensions is much larger than the other two, which are of the same order of magnitude (columns, thin beams, shafts, rings, etc.).

3. One of the dimensions is much smaller than the other two, which are of the same order of magnitude (thin plates, thin shells).

4. All three dimensions are of different orders of magnitude (thin, open-section beams).

Structural elements of the first category are not subject to instability. All other elements are. Typical stability problems associated with elements of the second category will be discussed in the subsequent chapters.

1.4 BRIEF HISTORICAL SKETCH

Structural elements that are subject to instability have been used for many centuries. Although their use is ancient, the first theoretical analysis of one such structural element (long column) was performed only a little over two hundred years ago. This first theoretical analysis is due to Leonhard Euler. Other men of the 18th and 19th centuries who are associated with theoretical and experimental investigations of stability problems are Lagrange, Considére, Bresse, M. Lévy, W. Fairbairn, A.G. Greenhill, F. Jasinski, Fr. Engesser, and G.H. Bryan. An excellent historical review is given by Timoshenko (Ref. 16) and Hoff (Ref. 17).

Bryan's work (Refs. 18 and 19) merits special attention because of his mathematical rigor and the novelty of the problems treated.

Among the existing texts on the subject, in addition to those cited so far, we should mention the books of Timoshenko and Gere (Ref. 20), Biezeno and Grammel (Ref. 21), Hoff (Ref. 22), Leipholz (Ref. 23), and Bleich (Ref. 24).

Significant contributions to the understanding of the concept of stability are those of Trefftz (Ref. 25), Pearson (Ref. 26), Koiter (Refs. 27 and 28), and Thompson (Ref. 29).

REFERENCES

1. WHITTAKER, E. T., *Analytical Dynamics*, Dover Publications, New York, 1944.

2. LANCZOS, C., *The Variational Principles of Mechanics*, University of Toronto Press, Toronto, 1960.

3. LANGHAAR, H. L., *Energy Methods in Applied Mechanics*, John Wiley & Sons, Inc., New York, 1962.

4. BOLOTIN, V. V., *Nonconservative Problems of the Theory of Elastic Stability*, edited by G. Herrmann (translated from the Russian), The Macmillan Co., New York, 1963.

5. HERRMANN, G., "Stability of the Equilibrium of Elastic Systems Subjected to Nonconservative Forces," *Applied Mechanics Review*, **20** 1967.

6. ZIEGLER, H., *Principles of Structural Stability*, Blaisdell Publishing Co., Waltham, Massachusetts, 1968.

7. LAGRANGE, J. L., *Mécanique Analytique*, Paris, 1788.

8. LIAPUNOV, A., "Problémé générale de la Stabilité du Mouvement," Traduit du Russe par E. Davaux. Annales de la Faculté des Sciences de Toulouse, Zéme serie, **9**, 1907. Reprinted by Princeton University Press, Princeton, 1952.

9. CHETAYEV, N. G., *Stability of Motion*, Translated from the second Russian edition by M. Nadler. Pergamon Press, London, 1961.

10. KRASOVSKII, N. N., *Stability of Motion*, Translated from the Russian by J. L. Brenner. Stanford University Press, Stanford, Calif., 1963.

11. LASALLE, J. P., and LEFSCHETZ, S., *Stability of Liapunov's Direct Method with Applications*, Academic Press, New York, 1961.

12. CHETAYEV, N. G., "Sur la Réciproque du théorème de Lagrange," *Comptes Rendues*, 1930.

13. POINCARÉ, H., "Sur le Equilibre d'une Masse Fluide Animée d'un Mouvement de Rotation," *Acta Mathematica*, Vol. 7, pp. 259–380, Stockholm, 1885.

14. FLÜGGE, W., *Handbook of Engineering Mechanics*, McGraw-Hill Book Company, New York, 1962, Chaps. 44 and 45.

15. BOLOTIN, V. V., *The Dynamic Stability of Elastic Systems*, Holden-Day, Inc., San Francisco, 1964.

16. TIMOSHENKO, S. P., *History of Strength of Materials*, McGraw-Hill Book Co., New York, 1953.

17. HOFF, N. J., "Buckling and Stability," *J. Royal Aero. Soc.*, Vol. 58, January, 1954.

18. BRYAN, G. H., "On the Stability of Elastic Systems," *Proc. Cambridge Phil. Soc.*, Vol. 6, p. 199, 1888.

19. BRYAN, G. H., "Buckling of Plates," *Proc. the London Math. Soc.*, Vol. 22, p. 54, 1891.

20. TIMOSHENKO, S. P., and GERE, J. M., *Theory of Elastic Stability*, McGraw-Hill Book Co., New York, 1961.

21. BIEZENO, C. B., and GRAMMEL, R., *Engineering Dynamics*, Vol. II, Part IV, Blackie and Son Ltd., London, 1956.

22. HOFF, N. J., *The Analysis of Structures*, John Wiley & Sons, Inc., New York, 1956.

23. LEIPHOLZ, H., *Stability Theory*, Academic Press, New York, 1970.

24. BLEICH, F., *Buckling Strength of Metal Structures*, McGraw-Hill Book Co., New York, 1952.

25. TREFFTZ, E., "Zur Theorie der Stabilität des Elastischen Gleichgewichts," *Zeitschr. f. angew. Math. u. Mech.*, Bd. 13, p. 160, 1933.

26. PEARSON, C. E., "General Theory of Elastic Stability," *Quarterly of Applied Mathematics*, Vol. 14, p. 133, 1956.

27. KOITER, W. T., "The Stability of Elastic Equilibrium," Thesis, Delft, 1945 (English translation NASA TT-F-10833, 1967).

28. KOITER, W. T., "Elastic Stability and Post-Buckling Behavior," in *Nonlinear Problems*, edited by R. E. Langer, University of Wisconsin Press, Madison, 1963, pp. 257–275.

29. THOMPSON, J. M. T., "Basic Principles in the General Theory of Elastic Stability," *J. of the Mechanics and Physics of Solids*, Vol. 11, p. 13, 1963.

2

MECHANICAL
STABILITY MODELS

Before undertaking the study of stability of elastic structures, the different methods available for understanding and obtaining critical conditions will be demonstrated through the use of simple mechanical models. The discussion will be limited to conservative systems. It is also intended to demonstrate the effect of geometric imperfections and load eccentricity on the response of the system. For a number of models, both the small-deflection (linear) and large-deflection (nonlinear) theories will be used for the sake of comparison. Finally, a comprehensive discussion of the different types of behaviors will be given to enhance understanding of buckling, critical conditions, and advantages or disadvantages of the approaches used.

2.1 MODEL A; A ONE-DEGREE-OF-FREEDOM MODEL

Consider a rigid bar of length l, hinged at one end, free at the other, and supported through a frictionless ring connected to a spring that can move only horizontally (see Fig. 2-1). The free end is loaded with a force P in the direction of the bar. It is assumed that the direction of the force remains unchanged. We may now ask: "Will the rigid bar remain in the upright position under the quasistatically applied load P?"

In trying to answer this question, we must consider all possible deflectional modes and study the stability of the system equilibrium. One possible deflectional mode allows rotation θ about the hinged end. In writing equilib-

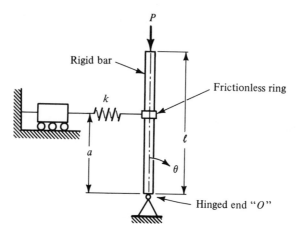

Figure 2-1. Geometry of Model *A*.

rium conditions for some θ positions, we could be interested in small θ as well as large θ values.

2.1-1 Small-θ Analysis

In the casual small-θ analysis, we make the usual assumption that θ is so small that $\theta \approx \sin \theta \approx \tan \theta$. With this restriction, we can only investigate the stability of the equilibrium configuration corresponding to $\theta = 0$. This type of investigation is sufficient to answer the posed question. The three approaches will be used separately.

1. The Classical or Equilibrium Method. The equilibrium equation corresponding to a deflected position is written under the assumption of small θ's (see Fig. 2-2). The expression for the moment about O is given by

$$M = -Pl\theta + (ka\theta)a$$

Since the bar is hinged at O, then

$$M = 0 \quad \text{or} \quad (Pl - ka^2)\theta = 0 \tag{1}$$

Thus a nontrivial solution exists if

$$Pl = ka^2$$

and the bifurcation point is located by

$$P = \frac{ka^2}{l} \quad \text{or} \quad P_{cr} = \frac{ka^2}{l}$$

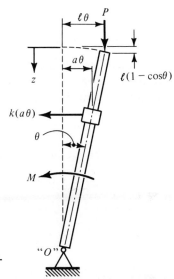

Figure 2-2. Small θ deflected position (Model A).

Figure 2-3 shows a plot of the load parameter $p = Pl/ka^2$ versus θ. Note that the bifurcation point is located at $p = 1$, and the $\theta \neq 0$ equilibrium positions are limited by the assumption of small θ.

2. Kinetic or Dynamic Approach. In this approach, we are interested in the character of the motion for small disturbances about the $\theta = 0$ position and at a constant P value. The equation of motion is given by

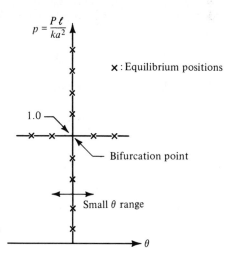

Figure 2-3. Load-deflection curve (Model A; small θ analysis).

$$I\ddot{\theta} + M = 0$$

or
$$\tag{2}$$
$$I\ddot{\theta} - (Pl - ka^2)\theta = 0$$

where dots above θ denote differentiation with respect to time, and I is the moment of inertia of the rigid bar about the hinged end O. It is easily seen from the differential equation that if

$$Pl - ka^2 < 0$$

the motion is oscillatory and the equilibrium is stable. If

$$Pl - ka^2 > 0$$

the motion is diverging and the equilibrium is unstable. If

$$Pl - ka^2 = 0$$

the motion can still be considered diverging (constant or linear with respect to time) and the equilibrium is unstable (neutrally stable).

Note that the frequency f is given by

$$f = \frac{1}{2\pi}\left(\frac{pl - ka^2}{-I}\right)^{1/2} \tag{3}$$

and at the critical condition $f = 0$, or

$$P_{cr} = \frac{ka^2}{l}$$

3. Energy Approach. Since the system is conservative, the externally applied force P can be derived from a potential. Thus (see Fig. 2-2)

$$U_p = P(z_0 - z)$$

and letting $z_0 = 0$, then $U_p = -Pz$ and

$$P = -\frac{dU_p}{dz} \tag{4}$$

On the other hand, the energy, U_i, stored in the system is given by

$$U_i = \tfrac{1}{2}k(a\theta)^2 \tag{5}$$

Thus the total potential U_T is given by

$$U_T = U_i + U_P$$
$$= -Pz + \tfrac{1}{2}k(a\theta)^2 \tag{6}$$

But since $z = l(1 - \cos \theta)$, then

$$U_T = -Pl(1 - \cos \theta) + \tfrac{1}{2}ka^2\theta^2 \tag{7}$$

For static equilibrium, the total potential must have a stationary value. Thus

$$\frac{dU_T}{d\theta} = 0$$

or

$$-Pl \sin \theta + ka^2\theta = 0$$

and since $\sin \theta \approx \theta$, then

$$(-Pl + ka^2)\theta = 0 \tag{8}$$

This is the same equilibrium equation we derived previously. Furthermore, if the second variation is positive definite, the static equilibrium is stable. If the second variation is negative definite, the static equilibrium is unstable; if it is zero, no conclusion can be drawn.

It is seen in this case that

$$\frac{d^2U_T}{d\theta^2} = ka^2 - Pl \tag{9}$$

Therefore, for $P < ka^2/l$ the static equilibrium positions ($\theta = 0$) are stable, while for $P > ka^2/l$ they are unstable. Thus, as before,

$$P_{cr} = \frac{ka^2}{l}$$

2.1-2 Large-θ Analysis

In this particular approach, the only limitation on θ is dictated from geometrical considerations. Note from Fig. 2-4 that $-\cos^{-1} a/l < \theta < \cos^{-1} a/l$. For θ values outside this range, the ring will fly off the rigid bar. As before, the three approaches shall be treated separately.

1. The Classical or Bifurcation Method. Since the ring is frictionless, the force R, normal to the rigid bar, is related to the spring force through the following expression (see Fig. 2-4):

$$k(a \tan \theta) = R \cos \theta$$

Then the moment about pin O is given by

$$M = -Pl \sin \theta + \frac{ka^2 \sin \theta}{\cos^3 \theta} \tag{10}$$

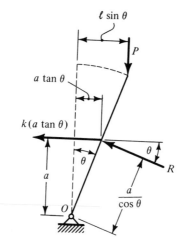

Figure 2-4. Geometry for large θ analysis (Model A).

For static equilibrium we require that $M = 0$. Thus, the equilibrium positions are characterized by the equation

$$\left(\frac{ka^2}{\cos^3 \theta} - Pl\right) \sin \theta = 0 \tag{11}$$

which implies that either

$$\theta = 0 \tag{12a}$$

or

$$\frac{Pl}{ka^2} = \sec^3 \theta \tag{12b}$$

It is clearly seen that a nontrivial solution ($\theta \neq 0$) can exist for $Pl/ka^2 > 1$ and a bifurcation point exists at $Pl/ka^2 = 1$ (see Fig. 2-5).

The answer to the original question is yes, and $P_{cr} = ka^2/l$.

2. Kinetic or Dynamic Approach.   In this approach, as before, we are interested in the character of the motion for small disturbances about the static equilibrium positions, keeping P constant. The equation of motion is given by

$$I\ddot{\theta} + M = 0$$

But

$$M = -Pl \sin \theta + \frac{ka^2 \sin \theta}{\cos^3 \theta} \tag{13}$$

If we denote the equilibrium positions by θ_0 and the disturbed positions by $\theta = (\theta_0 + \varphi)$, the Taylor-series expansion for the moment is given by

$$M(\theta_0 + \varphi) = M(\theta_0) + \varphi\left(\frac{dM}{d\theta}\right)_{\theta=\theta_0} + \frac{\varphi^2}{2!}\left(\frac{d^2M}{d\theta^2}\right)_{\theta=\theta_0} + \cdots \tag{14}$$

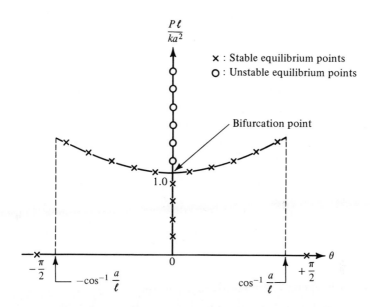

Figure 2-5. Load-deflection curve (Model A; large θ analysis).

At the static equilibrium positions, $M(\theta_0) = 0$. Through differentiation, we may write

$$\left(\frac{dM}{d\theta}\right)_{\theta_0} = \left(\frac{ka^2}{\cos^3\theta_0} - Pl\right)\cos\theta_0 + \sin\theta_0\left(3ka^2\frac{\sin\theta_0}{\cos^4\theta_0}\right) \qquad (15)$$

The equilibrium positions (see Fig. 2-4) are denoted by Eqs. (12). Thus, the equation of motion under the assumption of small disturbances for the equilibrium positions corresponding to $\theta_0 \neq 0$ is given by

$$I\ddot{\varphi} + 3ka^2\frac{\sin^2\theta_0}{\cos^4\theta_0}\varphi = 0 \qquad (16a)$$

and since

$$3ka^2\frac{\sin^2\theta_0}{\cos^4\theta_0} > 0$$

these equilibrium positions are stable. Note that $\theta_0 \neq 0$.

The equation of motion for the positions corresponding to $\theta_0 = 0$ is given by

$$I\ddot{\varphi} + (ka^2 - Pl)\varphi = 0 \qquad (16b)$$

If $Pl < ka^2$, the equilibrium is stable, while if $Pl < ka^2$, the equilibrium is unstable.

The equation of motion for the particular position corresponding to

$\theta_0 = 0$ and $Pl = ka^2$ is given below. We obtain this equation by taking more terms in the series expansion for M, Eq. (14).

$$I\ddot{\varphi} + \tfrac{3}{2}ka^2\varphi^3 = 0 \qquad (16c)$$

A study of this differential equation (see the following section, *Parenthesis*) indicates that the motion is stable. Although the equilibrium position $\theta_0 = 0$, $Pl = ka^2$ is stable, the answer to the original question is that the bar will not remain in the upright position, and the critical value of the load is given by

$$P_{cr} = \frac{ka^2}{l}$$

Parenthesis.   If the equation of motion of a nonlinear system is given by

$$\ddot{x} + k^2 x^{2n+1} = 0 \qquad (17)$$

where the dots above x denote differentiation with respect to time, k^2 is a positive number, and n is a positive integer, then the system is conservative and the position $x = 0$ is stable (see Refs. 1 through 4). What this implies physically is that, depending on the initial conditions, the total energy of the system is constant (sum of kinetic and potential energies is constant, thus the system is conservative) and the system performs nonlinear oscillations about the null position $x = 0$ within a bounded region enclosing the position $x = 0$.

The following computations will further clarify the above statements. Since

$$\ddot{x} = \frac{d\dot{x}}{dt} = \frac{d\dot{x}}{dx} \cdot \frac{dx}{dt} = \dot{x}\frac{d\dot{x}}{dx}$$

then Eq. (17) may now be written as

$$\dot{x}\frac{d\dot{x}}{dx} = -k^2 x^{2n+1}$$

or

$$\dot{x}\,d\dot{x} = -k^2 x^{2n+1}\,dx$$

If the initial conditions are denoted by $\dot{x}_0$ and x_0, then integration of this last equation yields

$$\tfrac{1}{2}[\dot{x}^2 - \dot{x}_0^2] = \frac{k^2}{2(n+1)}[x_0^{2(n+1)} - x^{2(n+1)}]$$

This equation expresses the law of conservation of energy. The left side denotes the change in kinetic energy and the right side denotes the change in potential energy. With reference to Eq. (16), x and $\dot{x}$ denote the size of the response to initial disturbances $\dot{x}_0$ and/or x_0. If we let the disturbance be x_0 only ($\dot{x}_0 \equiv 0$), then

$$\dot{x}^2 = \frac{k^2}{2(n+1)}[x_0^{2(n+1)} - x^{2(n+1)}]$$

which implies that the response is bounded.

3. Energy Approach. The total potential of the system is given by

$$U_T = -Pl(1 - \cos\theta) + \tfrac{1}{2}ka^2 \tan^2\theta \tag{18}$$

The static equilibrium positions are characterized by the equation

$$\frac{dU_T}{d\theta} = 0$$

or

$$\left(-Pl + \frac{ka^2}{\cos^3\theta}\right)\sin\theta = 0$$

Furthermore, the second variation is given by

$$\frac{d^2 U_T}{d\theta^2} = \left(\frac{ka^2}{\cos^3\theta} - Pl\right)\cos\theta + 3ka^2\frac{\sin^2\theta}{\cos^4\theta} \tag{19}$$

It is easily concluded that the static equilibrium positions characterized by $\theta \neq 0$ are stable. Similarly, the positions $\theta = 0$ for $Pl > ka^2$ are unstable. It can also be concluded, by considering higher variations, that the position denoted by $\theta_0 = 0$ and $Pl = ka^2$ is stable (see Chapter 1). The answer to the original question, though, still remains the same and

$$P_{cr} = \frac{ka^2}{l}$$

2.2 MODEL B; A ONE-DEGREE-OF-FREEDOM MODEL

Consider two rigid links pinned together and supported by hinges on rollers at the free ends (Fig. 2-6a). The system is supported at the middle hinge by a vertical linear spring and is acted upon by two collinear horizontal forces of equal intensity. The two links are initially horizontal. Can the system buckle? What is the critical load? To answer these questions, we may use small-deflection theory. The classical method and the energy method shall be used in this case.

1. The Classical or Bifurcation Method. Using casual small-deflection theory, we put the system into a deflected position (Fig. 2-6b) and write the equilibrium equations.

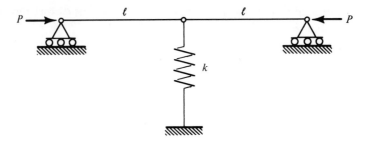

(a) Geometry; Model B

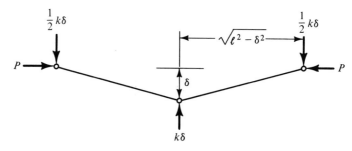

(b) Forces and displacements

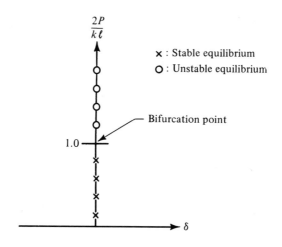

× : Stable equilibrium

○ : Unstable equilibrium

Bifurcation point

(c) Load-deflection curve

Figure 2-6. Model B.

Since the system is symmetric, the vertical reactions at the hinges are $k\delta/2$. Furthermore, the moment about the middle hinge must vanish. This requirement leads to the equilibrium equation

$$\left(P - \frac{kl}{2}\right)\delta = 0 \tag{20}$$

Thus, the equilibrium positions are defined by either $\delta = 0$ (trivial solution) or $P = kl/2$. In plotting $2P/kl$ versus δ, we notice that a bifurcation point exists at $2P/kl = 1$ (Fig. 2-6c) and

$$P_{cr} = \frac{kl}{2} \tag{21}$$

2. *Energy Method.* The total potential is the sum of the energy stored in the spring and the potential of the external forces. Thus

$$U_T = \frac{k\delta^2}{2} - 2P[l - \sqrt{l^2 - \delta^2}]$$

For static equilibrium $dU_T/d\delta = 0$ and

$$k\delta - 2P \cdot \frac{1}{2} \frac{2\delta}{(l^2 - \delta^2)^{1/2}} = 0 \tag{22}$$

which, under the assumption of $\delta^2 \ll l^2$, is identical to Eq. (20).

For the static equilibrium positions to be stable, the second variation must be positive definite, or

$$\frac{d^2 U_T}{d\delta^2} = k - \frac{2P}{(l^2 - \delta^2)^{1/2}} + \frac{2P\delta^2}{(l^2 - \delta^2)^{3/2}} \cong k - \frac{2P}{l} \tag{23}$$

Thus the equilibrium positions denoted by $\delta = 0$ and $P < kl/2$ are stable, and the critical load is given by Eq. (21).

2.3 MODEL C; A TWO-DEGREE-OF-FREEDOM MODEL

Consider the system shown in Fig. 2-7a, composed of three rigid bars of equal length hinged together as shown. The linear springs are of equal intensity. This is a two-degree-of-freedom system and it is acted upon by a horizontal force, P, applied quasistatically. We must determine whether or not the system will buckle and the critical value of the applied load. The load is assumed to remain horizontal.

1. *The Classical or Bifurcation Method.* In solving this problem, we will first use the classical method under the assumption of small deflections. Denoting by θ and φ the rotations about the support pins (see Fig. 2-7b)

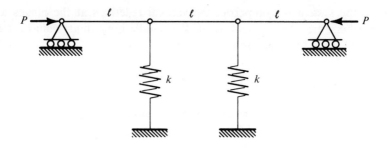

(a) Geometry

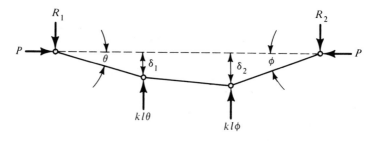

(b) Forces and displacements

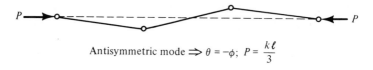

Antisymmetric mode $\Rightarrow \theta = -\phi$; $P = \dfrac{k\ell}{3}$

Symmetric mode $\Rightarrow \theta = \phi$; $P = k\ell$

(c) Mode shapes

Figure 2-7. Model *C.*

and by R_1 and R_2 the vertical reactions at the pins, we may write the following equilibrium equations for the deflected system:

$$\left.\begin{aligned}
3lR_1 &= 2kl^2\theta + kl^2\varphi\\
3lR_2 &= kl^2\theta + 2kl^2\varphi\\
R_1 l &= Pl\theta\\
R_2 l &= Pl\varphi
\end{aligned}\right\} \tag{24}$$

Elimination of R_1 and R_2 yields the following system of linear homogeneous algebraic equations:

$$\left.\begin{aligned}\left(P - \frac{2kl}{3}\right)\theta - \frac{kl\varphi}{3} &= 0\\ \frac{kl\theta}{3} - \left(P - \frac{2kl}{3}\right)\varphi &= 0\end{aligned}\right\} \tag{25}$$

The critical condition is derived if we require the existence of a nontrivial solution. This leads to the characteristic equation

$$\begin{vmatrix} \left(P - \dfrac{2kl}{3}\right) & -\dfrac{kl}{3} \\[2mm] \dfrac{kl}{3} & -\left(P - \dfrac{2kl}{3}\right) \end{vmatrix} = 0$$

from which

$$P = \left\{\begin{matrix} \dfrac{kl}{3} \\[2mm] kl \end{matrix}\right\} \tag{26}$$

Thus, there are two solutions (eigenvalues) corresponding to two modes of deformation (Fig. 2-7c):

$$P = \frac{kl}{3} \quad \text{and} \quad \varphi = -\theta$$

$$P = kl \quad \text{and} \quad \varphi = \theta$$

This shows that the smallest load corresponds to the antisymmetric mode.

2. *The Energy Method.* In Fig. 2-7b, the total potential for the system, which consists of the energy stored in the springs and the potential of the external forces, is given by

$$U_T = U_i + U_p = \tfrac{1}{2}kl^2\theta^2$$
$$+ \tfrac{1}{2}kl^2\varphi^2 - Pl[(1 - \cos\theta) + (1 - \cos\varphi) + 1 - \cos(\varphi - \theta)] \tag{27}$$

By assuming that the angles φ and θ can be made as small as desired, we may rewrite Eq. (27) as

$$U_T = \tfrac{1}{2}kl^2\theta^2 + \tfrac{1}{2}kl^2\varphi^2 - Pl(\theta^2 + \varphi^2 - \varphi\theta) \tag{28}$$

For static equilibrium, the total potential must be stationary; therefore

$$\frac{\partial U_T}{\partial \theta} = \frac{\partial U_T}{\partial \varphi} = 0 \tag{29}$$

which leads to the following equilibrium equations:

$$\begin{aligned}(kl^2 - 2Pl)\theta + Pl\varphi &= 0\\ Pl\theta + (kl^2 - 2Pl)\varphi &= 0\end{aligned} \right\}\tag{30}$$

The nontrivial solution is the same as the one obtained by the classical approach.

$$P = \frac{kl}{3}, \qquad P = kl \tag{31}$$

Study of the stability of the equilibrium positions characterized by $\theta = \varphi = 0$ for the entire range of values of P requires knowledge of the second variations

$$\frac{\partial^2 U_T}{\partial\theta^2} = kl^2 - 2Pl \tag{32}$$

$$\frac{\partial^2 U_T}{\partial\varphi^2} = kl^2 - 2Pl \tag{33}$$

$$\frac{\partial^2 U_T}{\partial\theta\,\partial\varphi} = Pl \tag{34}$$

The equilibrium positions are stable if and only if (see Chapter 1) both of the following inequalities are satisfied.

$$\frac{\partial^2 U}{\partial\theta^2} > 0$$

$$\frac{\partial^2 U_T}{\partial\theta^2} \cdot \frac{\partial^2 U_T}{\partial\varphi^2} > \left(\frac{\partial^2 U_T}{\partial\theta\,\partial\varphi}\right)^2 \tag{35}$$

In terms of the applied load and the structural geometry, these inequalities are

$$kl > 2P$$

$$(kl - P)\left(\frac{kl}{3} - P\right) > 0 \tag{36}$$

From these expressions, we see that equilibrium positions for which $P < kl/3$ are stable, while all equilibrium positions for which $P > kl/3$ are unstable. Therefore

$$P_{cr} = \frac{kl}{3}$$

2.4 MODEL *D*; A SNAPTHROUGH MODEL

In the analysis of this model, we will demonstrate the type of buckling known as snapthrough or oil-canning.

Consider two rigid bars of length *l* pinned together, with one end of the system pinned to an immovable support, and the other pinned to a linear horizontal spring. (See Fig. 2-8a.) The rigid bars make an angle α with the horizontal when the spring is unstretched and the system is loaded laterally through a force *P* applied quasistatically at the connection of the two rigid bars. As the load is increased quasistatically from zero, the spring will be compressed and the two bars will make an angle θ with the horizontal ($\theta < \alpha$). The question then arises whether it is possible for the system to snap-through toward the other side at some value of the applied load. In seeking the answer to this question, we will first use the equilibrium approach and then analyze the system by considering the character of the equilibrium positions. The latter will be accomplished through the energy approach.

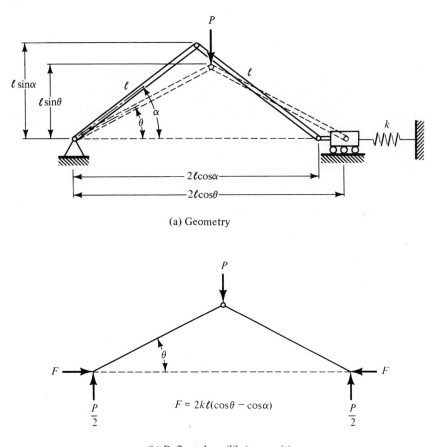

(a) Geometry

$$F = 2k\ell(\cos\theta - \cos\alpha)$$

(b) Deflected equilibrium position

Figure 2-8. Model *D*.

1. The Equilibrium Approach. Let the horizontal reaction of the spring be F. This force is equal to k times the compression in the spring (Fig. 2-8b), or

$$F = 2kl(\cos \theta - \cos \alpha) \qquad (37)$$

Furthermore, from symmetry the vertical reactions at the ends are $P/2$. Since no moment can be transferred through the middle joint, the equilibrium states are characterized by the following equation

$$\frac{Pl}{2} \cos \theta = Fl \sin \theta \qquad (38)$$

Use of Eq. (37) yields

$$\frac{P}{4kl} = \sin \theta - \cos \alpha \sin \theta \qquad (39)$$

Note that $-\pi/2 < \theta < \alpha < \pi/2$.

The equilibrium states, Eq. (39), are plotted in Fig. 2-9b. Note that loading starts at point A and it is increased quasistatically. When point B is reached, we see that with no appreciable change in the load the system will tend to snapthrough toward the CD portion of the curve. The load corresponding to position B is a critical one, and its magnitude may be obtained from the fact that

$$\frac{dP}{d\theta} = 0 \qquad (40)$$

Note that the right side of Eq. (39) is a continuous function of θ with continuous first derivatives.

If we denote by θ_B the angles corresponding to positions B and B', then

$$\theta_B = \pm \cos^{-1} (\cos \alpha)^{1/3} \qquad (41)$$

and

$$\left. \frac{P}{4kl} \right|_{cr} = |\sin \theta_B - \cos \alpha \tan \theta_B| \qquad (42)$$

2. Energy Approach. The total potential, U_T, for the system, which is equal to the potential of the external force and the energy stored in the spring, is given by

$$U_T = 2kl^2(\cos \theta - \cos \alpha)^2 - Pl(\sin \alpha - \sin \theta) \qquad (43)$$

Static equilibrium positions are characterized by the vanishing of the first variation of the total potential, or

$$\frac{dU_T}{d\theta} = 4kl^2(\cos \theta - \cos \alpha)(-\sin \theta) + Pl \cos \theta = 0$$

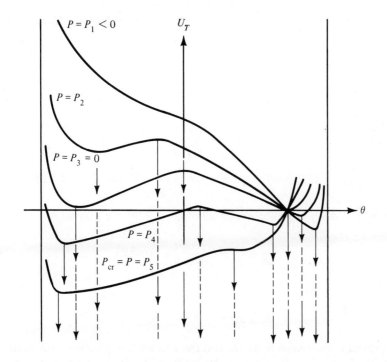

(a) Total potential curves at constant P

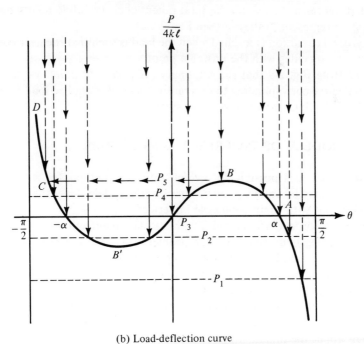

(b) Load-deflection curve

Figure 2-9. Critical conditions for model D.

This leads to the equilibrium equation

$$\frac{P}{4kl} = \sin\theta - \cos\alpha\tan\theta \tag{39}$$

The character of the equilibrium positions is governed by the second variation, or

$$\frac{d^2 U_T}{d\theta^2} = 4kl^2(\cos\theta - \cos\alpha)(-\cos\theta) + 4kl^2\sin^2\theta - Pl\sin\theta \tag{44}$$

Making use of the equilibrium condition, Eq. (39), we may write

$$\frac{d^2 U_T}{d\theta^2} = 4kl^2\left(\frac{\cos\alpha}{\cos\theta} - \cos^2\theta\right) \tag{45}$$

Thus in the region

$$-\cos^{-1}(\cos\alpha)^{1/3} < \theta < +\cos^{-1}(\cos\alpha)^{1/3}$$

the second derivative is negative and the equilibrium positions are unstable. Outside this region, the second derivative is positive and the equilibrium positions are stable. Thus, P_{cr} is given by Eq. (42). Note that points between B and B' represent "hills" on the total potential curve, while points outside this region represent "valleys." (See Fig. 2-9a.)

A critical condition is reached when the load is such that the near equilibrium point coincides with the unstable point.

Note from Fig. 2-9 that the stationary $(dU_T/d\theta = 0)$ points on the total potential curve corresponding to different values of the applied load make up the load-deflection curve (equilibrium states).

2.5 MODELS OF IMPERFECT GEOMETRIES

In many cases it is possible to predict critical conditions for a system of perfect geometry by studying the behavior of the system under the same load conditions but with slight geometric imperfections.

Consider, for instance, model B with a small imperfection δ_0 (Fig. 2-10a) when the spring is unstretched. The problem is to find the behavior of the imperfect system under the quasistatic application of the horizontal forces. Once this behavior has been established, the question arises whether or not we can predict the critical condition for the system of perfect geometry.

From the conditions of symmetry, the vertical reactions at the end pins are equal to $\frac{1}{2}k\delta$. The equilibrium condition is obtained if we require the moment about the middle hinge to vanish.

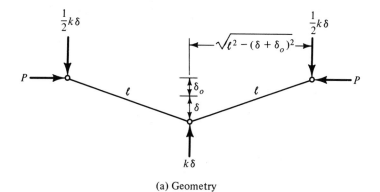

(a) Geometry

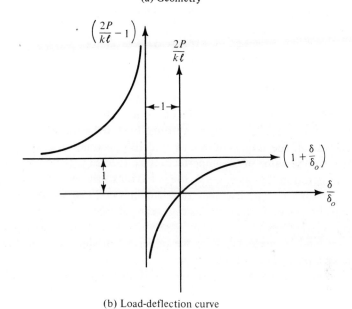

(b) Load-deflection curve

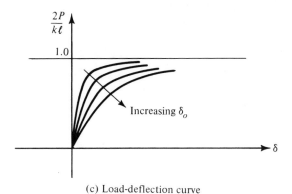

(c) Load-deflection curve

Figure 2-10. Model B with initial imperfection.

$$P(\delta + \delta_0) = \frac{k\delta}{2}\sqrt{l^2 - (\delta + \delta_0)^2}$$

$$\approx \frac{k\delta l}{2} \tag{46}$$

This equation can be written in the form

$$\left(P - \frac{kl}{2}\right)(\delta + \delta_0) = -\frac{kl}{2}\delta_0 \tag{47}$$

If we divide both sides by $(kl/2)\delta_0$, the equilibrium equation becomes

$$\left(\frac{2P}{kl} - 1\right)\left(1 + \frac{\delta}{\delta_0}\right) = -1 \tag{48}$$

This represents a hyperbola in the coordinate system of $(2P/kl - 1)$ and $(1 + \delta/\delta_0)$. (See Fig. 2-10b.) When a translation of axes is used, it appears that the load-deflection curve, in the coordinate system of $2P/kl$ and δ/δ_0, approaches the line $2P/kl = 1$ asymptotically. Furthermore, when $2P/kl$ is plotted versus δ, the single curve of Fig. 2-10b becomes a family of curves dependent on the value of the imperfection δ_0. (See Fig. 2-10c.) We see from this last illustration that as $\delta_0 \longrightarrow 0$, the behavior of the system is such that δ remains zero until $2P/kl$ becomes equal to unity. Thus, for $\delta_0 = 0$, $P_{cr} = kl/2$. This conclusion is the same as that reached when the system of perfect geometry was analyzed.

As a second example, consider the imperfect model shown in Fig. 2-11. Note that as the load eccentricity approaches zero, we have the corresponding perfect geometry model given in Problem 1 at the end of this chapter. For this particular problem, we want to find the effect of the eccentricity, e, on the critical load, P_{cr}. Once this effect is established, we can predict P_{cr} for the perfect configuration by letting the eccentricity approach zero. We will use the energy approach to solve the problem.

The total potential is given by

$$U_T = \frac{1}{2}ka^2 \sin^2\theta - Pl\left(1 + \cos\theta + \frac{e}{l}\sin\theta\right) \tag{49}$$

For equilibrium

$$\frac{\partial U_T}{\partial\theta} = 0 = ka^2\sin\theta\cos\theta - Pl\left(\sin\theta + \frac{e}{l}\cos\theta\right) \tag{50}$$

From this equation we obtain the load-deflection curve for a given load eccentricity e:

$$P = \frac{Pl}{ka^2} = \frac{\sin\theta}{\tan\theta + (e/l)} \tag{51}$$

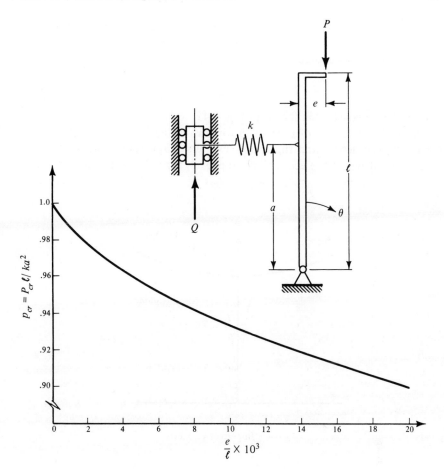

Figure 2-11. Effect of imperfection on the critical load.

Note that if e is replaced by $-e$ and θ by $-\theta$, we have the same load-deflection relation.

If we restrict the range of θ values to $0 < \theta < \pi/2$, we may study the second variation.

$$\frac{\partial^2 U_T}{\partial \theta^2} = \cos 2\theta - p\left(\cos \theta - \frac{e}{l}\sin \theta\right) \tag{52}$$

If we eliminate p, through Eq. (51), and use some well-known trigonometric identities, we finally obtain

$$\frac{\partial^2 U_T}{\partial \theta^2} = \frac{\cos^2 \theta}{\tan \theta + (e/l)}\left[-\tan^3 \theta + \frac{e}{l}\right]$$

Clearly, if $\tan^3 \theta < e/l$, the equilibrium positions are stable, and if $\tan^3 \theta > e/l$, the equilibrium positions are unstable. When $\tan^3 \theta = e/l$, $p = p_{cr}$, and substitution of this expression for θ into Eq. (51) yields

$$p_{cr} = \left[1 + \left(\frac{e}{l} \right)^{2/3} \right]^{-3/2} \tag{53}$$

A plot of p_{cr} versus e/l is shown in Fig. 2-11.

A (qualitative) plot of p versus θ for this model is given in Fig. 2-12(b) (imperfect geometry), and this model exhibits snapthrough buckling.

Finally, if we let the eccentricity approach zero, $P_{cr} = 1$ and

$$P_{cr} = \frac{ka^2}{l}$$

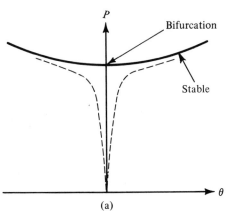

(a)

- - - - - : Imperfect geometry

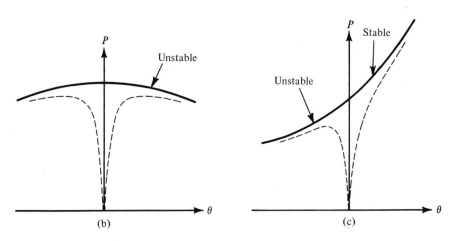

(b)   (c)

Figure 2-12. Possible load-deflection curves for bifurcation buckling.

2.6 DISCUSSION OF THE METHODS

After having considered these few mechanical models, certain observations must be discussed in order to enhance the understanding of the question of critical loads as well as the question of stability of elastic systems in general. In particular, attention is given to the relationship between the classical approach and the energy approach, which is completely equivalent to the dynamic approach for conservative systems (a proof of this is found in Ref. 7), and to the need for using large-deflection theories in certain problems.

We first noticed that, whenever the model exhibited a bifurcation point (models A and B), regardless of the approach used, the same result is obtained. On the other hand, when there is no bifurcation point (model D), the classical approach could only lead us to a load-deflection curve, and the criticality of the load at point B (see Fig. 2-9b) was explained as follows: If one wishes to increase the load any further, the system will visibly snap through toward a far equilibrium position. This argument of course implies deadweight-type of loading (prescription of the load rather than deflection), and it seems rather arbitrary. When the energy approach is used, it is very clear that the equilibrium positions between B and B' (Fig. 2-9b) are unstable, and therefore the load at B is critical because the slightest possible disturbance at this equilibrium position will make the system snap toward a far equilibrium position. In the absence of damping and assuming that the spring remains elastic, if the load at B is maintained, the system will simply oscillate (non-linearly) between θ_B and some angle past $-\alpha$. (See Fig. 2-9a.)

The second observation deals with the question of using large-deflection theories for predicting instability of perfect geometries (model A). It is clear that, when dealing with systems characterized by model D, large-deflection theory cannot be avoided. Therefore this question is directed to systems that exhibit bifurcational buckling (adjacent equilibrium position). From the examples considered, we may suspect that small-deflection theory suffices to predict critical loads. Since the analysis (models A, B and C) is based on the assumption that there are no imperfections in the geometry of the system, large-deflection theories are needed because they clearly indicate through the load-deflection curves (equilibrium positions) whether geometrical imperfections are likely to have a significant effect on the buckling of the real structure. Consider, for example, model A (Fig. 2-5). Small geometric imperfections have little effect on this system. This can be verified by the introduction of a small imperfection θ_0 and the use of a large-θ analysis on the imperfect system. The result is qualitatively shown in Fig. 2-12a. However, it can be demonstrated that small geometric imperfections can cause a dramatic reduction in the buckling load when the load-deflection curve is characterized by either Fig. 2-12b or Fig. 2-12c. Note that in all three cases (Fig. 2-12) the small-deflection theory can only predict the bifurcation load.

The stability of structures immediately after buckling (bifurcation) was first investigated systematically by Koiter (Ref. 6), and alternative formulations of the general theory have subsequently been given by Thompson (Refs. 8 and 9) and by Sewell (Ref. 10). Thompson (Ref. 9) and Pope (Ref. 11) show that the derivative $dp/d\theta$ at the bifurcation can be calculated exactly, for complicated elastic systems, by a finite-deformation analysis of the Rayleigh-Ritz type. Some remarks on Koiter's theory are presented in Chapter 5.

PROBLEMS

1. Analyze the system shown using large-deflection theory. Give the load-deflection curve and the critical load.
   (a) Use the classical approach.
   (b) Use the kinetic approach.
   (c) Use the energy approach.

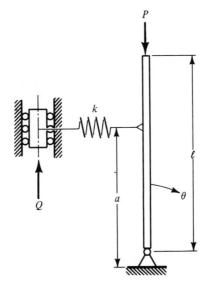

Figure P2-1.

2. A uniform disc can rotate freely about O, except that it is restrained by a rotational spring giving a restoring couple $\alpha\theta$ for angular displacement θ. A weight W is attached at radius a and vertically above O.
   (a) Show that a stable tilted position, θ_0, of equilibrium is possible when $W > \alpha/a$.
   (b) Show that when $W > \alpha/a$, the frequency of small oscillations about the position of stable equilibrium is

$$\frac{1}{2\pi}\sqrt{\frac{\alpha - Wa\cos\theta_0}{I}}$$

   where I is the moment of inertia (including Wa^2/g).

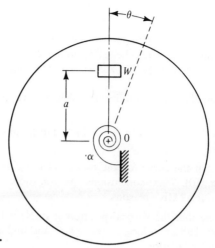

Figure P2-2.

(c) Show that when $W = \alpha/a$, the differential equation for small oscillations is

$$I\frac{d^2\theta}{dt^2} + \frac{1}{6}\alpha\theta^3 = 0$$

3. In the mechanism shown, a light stiff rod is pinned at O. There is no friction. P remains vertical if the bar tilts.
   (a) By using any method, find the $P - \theta$ relation for equilibrium positions, and plot the curve.

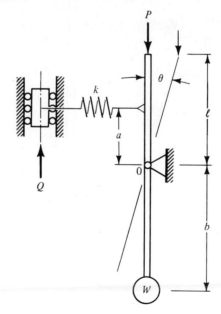

Figure P2-3.

(b) Discuss the stability or instability of all the equilibrium positions in the entire practical range of θ values.

4. In the coplanar system shown, the initially vertical rod is rigid. The block to which the spring is attached slides in the inclined guide and is controlled so that the spring is always horizontal. All parts have negligible mass except the weight W.

(a) Show that tilted equilibrium positions are characterized by

$$W = \frac{ka^2}{l}\left(1 - \tan\beta \tan\frac{\theta}{2}\right)\cos\theta$$

(b) Sketch the curve for the two cases $\tan\beta$ small (e.g., $\frac{1}{20}$) and $\tan\beta$ large (e.g., 10). What conclusions can you draw as to the stability of the tilted position? Give reasons.

(c) Show that the vertical position is stable with respect to sufficiently small disturbances so long as $Wl < ka^2$, and find a formula for the frequency of small oscillations.

(d) Show that when $Wl = ka^2$, the beginning of the motion from $\theta = 0$ following a slight disturbance will be governed by the equation

$$\frac{Wl^2}{g}\ddot{\theta} - \frac{1}{2}ka^2\tan\beta\cdot\theta^2 = 0$$

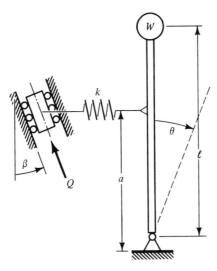

Figure P2-4.

5. Analyze model C by assuming that the lengths of the rigid bars and the spring stiffnesses are unequal. Let these lengths be l_1, l_2, and l_1 starting from the left. Let the spring constants be k for both.

(a) Use the classical approach.

(b) Use the energy approach.

6. Consider the rigid bar shown with an initial rotation θ_0 and initial stretch c of the spring. Use small-deflection theory, and through a complete analysis of the behavior of the imperfect system, predict critical conditions for the perfect system ($\theta_0 = c = 0$).

7. Repeat Problem 6 assuming that the initial stretch, c, is zero.

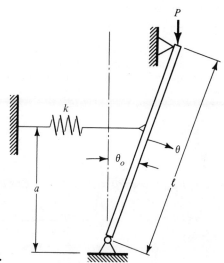

Figure P2-6 and 7.

8. Two rigid bars are connected by rotational springs to each other and to the support at C. Find P_{cr}, assuming that the load remains vertical.

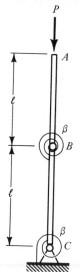

Figure P2-8.

9. Find the critical condition for model D through the kinetic approach. *Hint:* Consider the left leg as a free body and study its motion about the immovable support.

10. Consider the model shown loaded by a vertical force, P, applied quasistatically. Establish critical conditions for the system (for $C = 0$).
    (a) Use the equilibrium approach.
    (b) Use the energy approach.

11. Repeat Problem 10 assuming C is constant.

12. Repeat Problem 10 assuming $C = A + B \sin \theta + D \sin^2 \theta$ (nonlinear spring). Note that the numerical work involved is complicated and a computer program is needed as well as knowledge of the values of the different parameters.

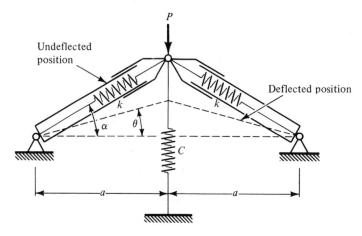

Figure P2-10, 11, and 12.

REFERENCES

1. STOKER, J. J., *Nonlinear Vibrations*, Interscience Publishers, Inc., New York, 1950.

2. MALKIN, I. G., *Theory of Stability of Motion*, A.E.C. Translation 3352, Dept. of Commerce, U.S.A., 1958.

3. LaSALLE, J. P., and LEFSCHETZ, S. *Stability of Liapunov's Direct Method With Applications*, Academic Press, New York, 1961.

4. KRASOVSKII, N. N., *Stability of Motion*, translated from the Russian by J. L. Brenner, Stanford University Press, Stanford, Calif., 1963.

5. HAYASHI, C., *Nonlinear Oscillations in Physical Systems*, McGraw-Hill Book Co., New York, 1964.

6. KOITER, W. T., "The Stability of Elastic Equilibrium," Thesis, Delft, 1945 (English translation NASA TT-F-10833, 1967).

7. WHITTAKER, E. T., *Analytical Dynamics*, Dover Publications, New York, 1944.

8. THOMPSON, J. M. T., "Basic Principles in the General Theory of Elastic Stability," *Journal of the Mechanics and Physics of Solids*, Vol. 11, p. 13, 1963.

9. THOMPSON, J. M. T., "Eigenvalue Branching Configurations and the Rayleigh-Ritz Procedure," *Quarterly of Appl. Math*, Vol. 22, p. 244, 1964.

10. SEWELL, M. J., "On the Connexion Between Stability and the Shape of the Equilibrium Surface," *Journal of the Mechanics and Physics of Solids*, Vol. 14, p. 203, 1966.

11. POPE, G. G., "On the Bifurcational Buckling of Elastic Beams, Plates and Shallow Shells," *The Aeronautical Quarterly*, p. 20, Feb., 1968.

3

ELASTIC BUCKLING
OF COLUMNS

In this chapter, the problem of elastic buckling of bars will be studied using the approach discussed in Chapter 1 and demonstrated in Chapter 2. To accomplish this, we will derive the equations governing equilibrium for structural elements of class 2 in Chapter 1 (Section 1.3) along with the proper boundary conditions. This derivation is based on the Euler-Bernoulli assumptions, listed below, and principle of the stationary value of the total potential (Appendix).

In analyzing slender rods and beams, we make the following basic engineering assumptions:

1. The material of the element is homogeneous and isotropic.
2. Plane sections remain plane after bending.
3. The stress-strain curve is identical in tension and compression.
4. No local type of instability will occur.
5. The effect of transverse shear is negligible.
6. No appreciable initial curvature exists.
7. The loads and the bending moments act in a plane passing through a principal axis of inertia of the cross section.
8. Hooke's law holds.
9. The deflections are small as compared to the cross-sectional dimensions.

In addition, the loads are assumed to be coplanar, applied quasistatically, and are either axial or transverse. The transverse loads include distributed loads, $q(x)$, concentrated loads, P_i, and concentrated couples, C_j. Finally, the ends of the structures are supported in such a way that primary degrees of freedom (translation and rotation as a rigid body) are constrained. Before proceeding with the derivation of the equilibrium equations and boundary conditions, it is desirable to define and discuss the properties of some special functions.

3.1 SPECIAL FUNCTIONS

The following special functions and their properties will be used in the development of the theory of slender rods and beams. See Fig. 3-1 for their graphical representation.

1. Macauley's Bracket

$$[x - x_i] = \begin{cases} 0 & \text{for } x < x_i \\ x - x_i & \text{for } x > x_i \end{cases} \tag{1}$$

2. Unit Step Function

$$I(x - x_i) = \begin{cases} 0 & \text{for } x < x_i \\ 1 & \text{for } x \geq x_i \end{cases} \tag{2}$$

Note that

$$[x - x_i] = (x - x_i)I(x - x_i)$$

and similarly

$$[x - x_i]^2 = (x - x_i)^2 I(x - x_i)$$

3. Dirac δ-Function (Ref. 1). The Dirac δ-function in all applications is considered as a result of a limiting process which involves a function $\delta(x, \epsilon)$ subject to the following conditions:

$$\delta(x, \epsilon) \geq 0 \quad \text{for} \quad -\infty \leq x \leq \infty \quad \text{and} \quad 0 < \epsilon < \infty$$

$$\int_{-\infty}^{\infty} \delta(x, \epsilon) \, dx = 1 \quad \text{for} \quad 0 < \epsilon < \infty$$

An example of such a function is the following:

$$\delta(x - x_i) = \begin{cases} 0 & \text{if } x < x_i - \epsilon \\ \dfrac{1}{2\epsilon} & \text{if } x_i - \epsilon \leq x \leq x_i + \epsilon \\ 0 & \text{if } x > x_i + \epsilon \end{cases} \tag{3}$$

Note that

$$\int_{-\infty}^{\infty} f(x) \, \delta(x - x_i) \, dx = f(x_i)$$

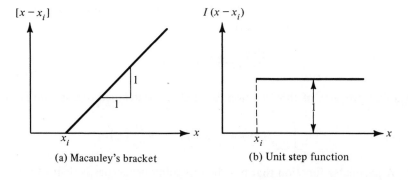

(a) Macauley's bracket (b) Unit step function

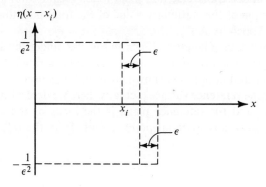

(c) Dirac δ-function

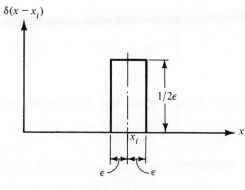

(d) The doublet function

Figure 3-1. Special functions.

4. *The Doublet Function.* Let this function be denoted by $\eta(x - x_i)$. This is a special function such that

$$\frac{d\delta(x - x_i)}{dx} = \eta(x - x_i)$$

Another property of this function (see Ref. 2 for detailed discussion) is that

$$\int_{-\infty}^{+\infty} f(x)\eta(x - x_i)\,dx = -\frac{df}{dx}(x_i)$$

A particular function that has the foregoing properties is defined as

$$\eta(x - x_i) = \begin{cases} 0 & x < x_i \\ \dfrac{1}{\epsilon^2} & x_i < x < x_i + \epsilon \\ 0 & x = x_i + \epsilon \\ -\dfrac{1}{\epsilon^2} & x_i + \epsilon < x < x_i + 2\epsilon \\ 0 & x > x_i + 2\epsilon \end{cases} \tag{4}$$

3.2 BEAM THEORY

The equilibrium equations and proper boundary conditions for an initially straight beam under transverse and axial loads (beam-column) are derived using the principle of the stationary value of the total potential. (See Part II of Ref. 3 and Appendix A.)

Consider the beam of length l, shown in Fig. 3-2, under the action of a distributed local $q(x)$, n concentrated forces, P_i, m concentrated couples, C_j, and boundary forces and couples as shown. If u and w denote the displacement components of the reference surface (actually, here we deal with a two-dimensional problem, and the reference plane is the locus of the centroids), the extensional strain of any material point, z units from the reference surface, is given by

$$\epsilon_{xx} = \epsilon_{xx}^0 + z\kappa_{xx} \tag{5}$$

where ϵ_{xx}^0 is the extensional strain on the reference plane (average strain) and κ_{xx} is the change in curvature of the reference plane.

The first-order nonlinear strain-displacement relation is given by

$$\epsilon_{xx}^0 = u_{,x} + \tfrac{1}{2}w_{,x}^2 \tag{6}$$

where the comma denotes differentiation with respect to coordinate x. u and w are displacement components of the reference plane.

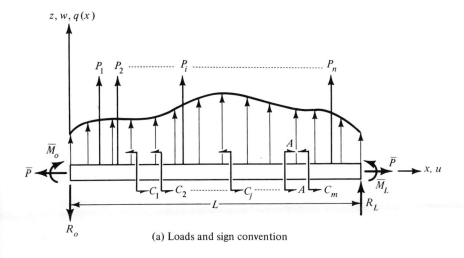

(a) Loads and sign convention

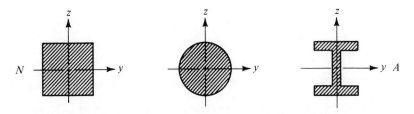

y and z: Principal centroidal axes
NA: Neutral axis

(b) Section A-A

Figure 3 2. Beam geometry and sign convention.

The curvature for the reference plane is approximated by

$$\kappa_{xx} = -w_{,xx} \tag{7}$$

In addition, the mathematical expression of Hooke's law is

$$\sigma_{xx} = E\epsilon_{xx} \tag{8}$$

If U_i and U_p denote the strain energy and potential of external forces, respectively, and U_T, the total potential of the system, then

$$U_i + U_p = U_T$$

Use of the principle of the stationary value of the total potential leads to

$$\delta_\epsilon U_T = \delta_\epsilon U_i + \delta_\epsilon U_p = 0 \tag{9}$$

where ϵ denotes variations with respect to strains and displacements.

Since the variation of the strain energy (see Part II of Ref. 3) is given by

$$\delta_\epsilon U_i = \int_V \sigma_{xx} \delta\epsilon_{xx} \, dV$$

$$= \int_0^L \int_A E[\epsilon_{xx}^0 + z\kappa_{xx}][\delta\epsilon_{xx}^0 + z\delta\kappa_{xx}] \, dA \, dx \tag{10}$$

and since ϵ_{xx}^0, κ_{xx}, and their variations are only functions of x (note that x is a centroidal axis), then

$$\int_A z \, dA = 0$$

and

$$\delta_\epsilon U_i = \int_0^L [P\delta\epsilon_{xx}^0 + EI\kappa_{xx}\delta\kappa_{xx}] \, dx$$

where

$$P = E\epsilon_{xx}^0 \int_A dA = EA\epsilon_{xx}^0$$

and

$$I = \int_A z^2 \, dA$$

Next, replacing the change in curvature and the variations in strain and change in curvature by the displacement components and their variations, we obtain

$$\delta_\epsilon U_i = \int_0^L [P(\delta u_{,x} + w_{,x}\delta w_{,x}) + EI w_{,xx}\delta w_{,xx}] \, dx \tag{11}$$

Note that

$$\delta_\epsilon(\tfrac{1}{2}w_{,x}^2) = \tfrac{1}{2}[(w_{,x} + \delta w_{,x})^2 - w_{,x}^2] = \tfrac{1}{2}[2w_{,x}\delta w_{,x} + (\delta w_{,x})^2]$$

and neglecting higher-order terms (assuming small changes),

$$\delta_\epsilon(\tfrac{1}{2}w_{,x}^2) = w_{,x}\delta w_{,x}$$

Integration by parts of Eq. (11) yields

$$\delta_\epsilon U_i = P\delta u \Big|_0^L + Pw_{,x}\delta w \Big|_0^L + EI w_{,xx}\delta w_{,x} \Big|_0^L - (EI w_{,xx})_{,x}\delta w \Big|_0^L$$

$$+ \int^L [-P_{,x}\delta u - (Pw_{,x})\delta w_{,x} + (EI w_{,xx})_{,xx}\delta w] \, dx \tag{12}$$

Similarly, the variation in the potential of the external forces is given by

$$\delta_\epsilon U_p = -\int_0^L q\delta w\, dx - \sum_{i=1}^n P_i\delta w(x_i) - \sum_{j=1}^m C_j\delta w_{,x}(x_j)$$
$$- \bar{P}u(L) + \bar{P}u(0) + M_0\delta w_{,x}(0) - M_L\delta w_{,x}(L) + R_0\delta w(0) - R_L\delta w(L)$$
$$(13)$$

If we introduce the special functions $\delta(x - x_i)$ and $\eta(x - x_i)$, the expression for the variation of the potential of the external forces becomes

$$\delta_\epsilon U_p = -\int_0^L \left[q + \sum_{i=1}^n P_i\delta(x - x_j) - \sum_{j=1}^m C_j\eta(x - x_j) \right]\delta w\, dx$$
$$- (\bar{P}\delta u)\Big|_0^L - (\bar{M}\delta w_{,x})\Big|_0^L - (R\delta w)\Big|_0^L \qquad (14)$$

Substitution of Eqs. (12) and (14) into Eq. (9) finally yields

$$\delta_\epsilon U_T = \int_0^L \left[-P_{,x}\delta u + \left\{ (EIw_{,xx})_{,xx} - (Pw_{,x})_{,x} - q - \sum_{i=1}^n P_i\delta(x - x_i) \right. \right.$$
$$\left. \left. + \sum_{j=1}^m C_j\eta(x - x_j) \right\}\delta w \right]dx + (P - \bar{P})\delta u\Big|_0^L$$
$$+ [Pw_{,x} - (EIw_{,xx})_{,x} - R]\delta w\Big|_0^L + [EIw_{,xx} - \bar{M}]\delta w_{,x}\Big|_0^L = 0$$
$$(15)$$

The identical satisfaction of Eq. (15) (since δu and δw are arbitrary displacement functions) leads to the governing differential equations and the proper boundary conditions.

The differential equations are

$$P_{,x} = 0$$
$$(EIw_{,xx})_{,xx} - Pw_{,xx} = q + \sum_{i=1}^n P_i\delta(x - x_i) - \sum_{j=1}^m C_j\eta(x - x_j) \qquad (16)$$

The proper boundary conditions are given by

Either	Or
$+P = \bar{P}$	$\delta u = 0 \Rightarrow u = \bar{u}$
$+\bar{P}w_{,x} - (EIw_{,xx})_{,x} = R$	$\delta w = 0 \Rightarrow w = 0$
$EIw_{,xx} = \bar{M}$	$\delta w_{,x} = 0 \Rightarrow w_{,x} = 0$

It is clearly shown above that, at the boundaries $(x = 0, L)$, we may prescribe either the forces and moments or the displacements and rotations, but not both.

Examples:

A free edge with no moment or shear force applied is characterized by

$$+\bar{P}w_{,x} - (EIw_{,xx})_{,x} = 0, \quad EIw_{,xx} = 0, \quad \text{and} \quad +P = \bar{P} \quad \text{or} \quad u = \bar{u} \quad (17a)$$

A simply supported edge is characterized by

$$w = 0, \quad EIw_{,xx} = 0, \quad \text{and} \quad +P = \bar{P} \quad \text{or} \quad u = \bar{u} \quad (17b)$$

Finally, a clamped edge is characterized by

$$w = 0, \quad w_{,x} = 0, \quad \text{and} \quad +P = \bar{P} \quad \text{or} \quad u = \bar{u} \quad (17c)$$

Note that the first of Eqs. (16) implies that $P = \text{constant}$, and from the boundary condition, this constant is equal to $\bar{P}$. In the case where the end shortening is prescribed $(\bar{u})$, there is a $P = \text{constant}$ corresponding to each value of $\bar{u}$.

3.3 BUCKLING OF COLUMNS

When a bar is initially straight and of perfect geometry and it is subjected to the action of a compressive force without eccentricity, then it is called an *ideal column*. When the load is applied quasistatically, the column is simply compressed but remains straight. We then need to know if the column will remain straight no matter what the level of the applied force is. To determine this, we seek nontrivial solutions $(w \neq 0)$ for the equations governing the bending [Eqs. (16) with $q = 0$, $P_i = 0$, and $C_j = 0$] of this column under an axial compressive load $(-\bar{P})$ and subject to the particular set of boundary conditions. Note that in deriving the governing differential equations, it was assumed that the applied compressive load remained parallel to its original direction and there was no eccentricity in either the geometry or the applied load. Thus, the problem has been reduced to an eigen-boundary-value problem.

3.3-1 Solution

In this case the solution of the problem will be discussed for a number of boundary conditions. It will be shown that the manner in which the column is supported at the two ends affects the critical load considerably. This approach to the problem is known as the classical, equilibrium, or bifurcation approach. In addition, the other approaches (dynamic and energy) will be demonstrated.

1. Simply Supported Ideal Column. The mathematical formulation of this problem is given below.

$$\text{D.E.} \qquad (EIw_{,xx})_{,xx} + \bar{P}w_{,xx} = 0 \qquad (18a)$$

$$\text{B.C.'s} \qquad w(0) = w(L) = 0$$
$$w_{,xx}(0) = w_{,xx}(L) = 0 \qquad (18b)$$

Assuming that the bending stiffness (*EI*) of the column is constant and introducing the parameter $k^2 = \bar{P}/EI$ allows us to write the governing differential equation in the following form

$$w_{,xxxx} + k^2 w_{,xx} = 0 \qquad (18c)$$

The general solution of this equation is given by

$$w = A_1 \sin kx + A_2 \cos kx + A_3 x + A_4 \qquad (19)$$

This solution must satisfy the prescribed boundary conditions. This requirement leads to four linear homogeneous algebraic equations in the four constants A_i. A nontrivial solution then exists if all four constants are not identically equal to zero. This can happen only if the determinant of the coefficients of the A_i's vanishes or

$$
\begin{array}{cccc}
A_1 & A_2 & A_3 & A_4 \\
\end{array}
$$
$$
\begin{vmatrix}
0 & 1 & 0 & 1 \\
\sin kL & \cos kL & L & 1 \\
0 & -k^2 & 0 & 0 \\
-k^2 \sin kL & -k^2 \cos kL & 0 & 0
\end{vmatrix} = 0 \qquad (20)
$$

The expansion of this determinant leads to

$$\sin kL = 0$$

The solution of this equation is

$$kL = n\pi \qquad n = 1, 2, \ldots$$

or

$$\bar{P} = \frac{n^2 \pi^2 EI}{L^2}$$

and the smallest of these corresponds to $n = 1$. Thus

$$P_{cr} = \frac{\pi^2 EI}{L^2}$$

This is known as the Euler equation because the problem was first solved by Leonhard Euler (see Ref. 4).

Note that, if A denotes the cross-sectional area of the column and p is the radius of gyration of the cross-sectional area, the critical stress is given by

$$\sigma_{cr} = \frac{\pi^2 EI}{AL^2} = \frac{\pi^2 p^2 E}{L^2} = \frac{\pi^2 E}{(L/p)^2} \qquad (21a)$$

and the corresponding strain is

$$\epsilon_{cr} = \left(\frac{\pi p}{L}\right)^2 \qquad (21b)$$

The displacement function corresponding to $n = 1$ is $w = A_1 \sin \pi x/L$.

2. Clamped Ideal Column.   For this particular problem, the mathematical formulation is given below

$$\text{D.E.} \qquad w_{,xxxx} + k^2 w_{,xx} = 0$$
$$\text{B.C.'s} \qquad w(0) = w(L) = 0 \qquad (22)$$
$$w_{,x}(0) = w_{,x}(L) = 0$$

The solution is given by Eq. (19) and it must satisfy the prescribed boundary conditions. The characteristic equation for this case is given by

$$\begin{vmatrix} 0 & 1 & 0 & 1 \\ \sin kL & \cos kL & L & 1 \\ k & 0 & 1 & 0 \\ k\cos kL & -k\sin kL & 1 & 0 \end{vmatrix} = 0 \qquad (23)$$

Expansion of this determinant yields the following equation

$$2(\cos kL - 1) + kL \sin kL = 0 \qquad (24)$$

Since

$$\cos kL - 1 = -2\sin^2 \frac{kL}{2}$$

and

$$\sin kL = 2\sin \frac{kL}{2} \cos \frac{kL}{2}$$

then Eq. (24) becomes

$$\sin \frac{kL}{2} \left(\frac{kL}{2} \cos \frac{kL}{2} - \sin \frac{kL}{2} \right) = 0$$

Then either

$$\sin \frac{kL}{2} = 0$$

or

$$\frac{kL}{2} \cos \frac{kL}{2} = \sin \frac{kL}{2}$$

The first of these solutions leads to

$$\bar{P} = \frac{4n^2\pi^2 EI}{L^2} \qquad n = 1, 2, \ldots$$

from which

$$P_{cr} = \frac{4\pi^2 EI}{L^2} \tag{25a}$$

and

$$\sigma_{cr} = \frac{4\pi^2 E}{(L/\rho)^2} \tag{25b}$$

$$\epsilon_{cr} = 4 \left(\frac{\pi\rho}{L} \right)^2$$

The second of the solutions leads to $P_{cr} > 4\pi^2 EI/L^2$. The displacement function corresponding to $n = 1$ is

$$w = A_2 \left(\cos \frac{2\pi x}{L} - 1 \right)$$

3. Ideal Column with One End Clamped and Other Free. For this case the boundary conditions are (assuming that the fixed end is at $x = 0$)

$$\left. \begin{array}{r} w(0) = w_{,x}(0) = 0 \\ w_{,xx}(L) = 0 \\ k^2 w_{,x}(L) + w_{,xxx}(L) = 0 \end{array} \right\} \tag{26}$$

The solution is still given by Eq. (19) and the characteristic equation is

$$\begin{vmatrix} 0 & 1 & 0 & 1 \\ k & 0 & 1 & 0 \\ -k^2 \sin kL & -k^2 \cos kL & 0 & 0 \\ 0 & 0 & k^2 & 0 \end{vmatrix} = 0 \tag{27}$$

which expanded results into the following equation

$$-k^5 \cos kL = 0 \left.\vphantom{\begin{matrix}a\\b\end{matrix}}\right\}$$

or

$$\cos kL = 0 \qquad (28)$$

This equation leads to the following result

$$\bar{P} = \left(\frac{2m-1}{2}\right)^2 \frac{\pi^2 EI}{L^2} \qquad m = 1, 2, \ldots$$

and

$$P_{cr} = \frac{\pi^2 EI}{4L^2}$$

$$\sigma_{cr} = \frac{\pi^2 E}{4(L/\rho)^2} \qquad (29)$$

$$\epsilon_{cr} = \frac{1}{4}\left(\frac{\pi\rho}{L}\right)^2$$

The displacement function for this case is given by

$$w = A_2\left(\cos\frac{\pi x}{2L} - 1\right)$$

3.3-2 Reduction of the Order of the Differential Equation

If the moment and shear are prescribed at $x = 0$, then it is possible to reduce the order of the governing differential equation from four to two.

Starting with Eq. (18a) under the assumption of constant flexural stiffness, two consecutive integrations yield the following equations

$$EIw_{,xxx} + \bar{P}w_{,x} = B_1$$

and

$$EIw_{,xx} + \bar{P}w = B_1 x + B_2$$

If, in addition, the normal displacement w is measured from the left end, then $w(0) = 0$.

Then the constants B_1 and B_2 can be evaluated from the known moment and shear and (see Fig. 3-2; R_1 opposite to positive w).

$$B_1 = -R_0$$
$$B_2 = +\bar{M}_0$$

Thus the governing differential equation reduces to

$$EIw_{,xx} + \bar{P}w = -R_0 x + \bar{M}_0 \qquad (30)$$

Note that, for the case of simply supported ideal columns, $\bar{R}_0 = \bar{M}_0 = 0$ and the equation becomes

$$EIw_{,xx} + \bar{P}w = 0 \tag{31}$$

3.3-3 Effective Slenderness Ratio

We have seen from the previous sections that the critical load of a compressed ideal column is affected by the boundary conditions. For all possible boundary conditions, the critical load may be expressed by

$$P_{cr} = C\frac{\pi^2 EI}{L^2} \tag{32}$$

where C is a constant that depends on the boundary conditions and is called the *end fixity factor*. Thus for both ends simply supported, $C = 1$; for one end fixed and the other free, $C = \frac{1}{4}$; and so on (see Table 3-1).

Similarly, the corresponding critical stress is given by

$$\sigma_{cr} = C\frac{\pi^2 E}{(L/\rho)^2} \tag{33}$$

where the parameter L/ρ is known as the slenderness ratio of the ideal column. For the case of simply supported ends, the end fixity factor is equal to unity. By defining a new parameter L' through

$$L' = \frac{L}{\sqrt{C}} \tag{34}$$

we may use the simply-supported-ends equation, for all cases of boundary conditions, through the equivalent slenderness ratio L'/ρ. Thus

$$\sigma_{cr} = \frac{\pi^2 E}{(L'/\rho)^2} \tag{35}$$

Note that a single curve of σ_{cr} plotted versus L'/ρ represents critical loads for all possible geometries and boundary conditions of ideal columns (see Fig. 3-3).

Table 3-1. END FIXITY FACTORS

Boundary Conditions	C	L'
Both ends simply supported	1	L
One end fixed, the other free	$\frac{1}{4}$	$2L$
Both ends fixed	4	$0.5L$
One end fixed, the other simply supported	$\left(\frac{4.493}{\pi}\right)^2$	$0.699L$

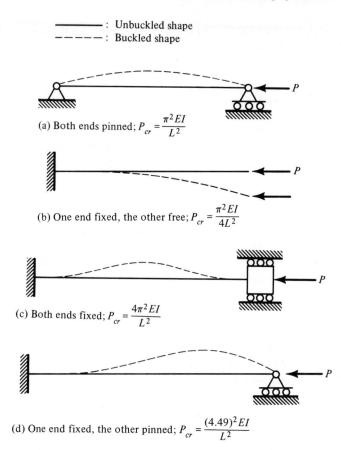

——————— : Unbuckled shape

— — — — — : Buckled shape

(a) Both ends pinned; $P_{cr} = \dfrac{\pi^2 EI}{L^2}$

(b) One end fixed, the other free; $P_{cr} = \dfrac{\pi^2 EI}{4L^2}$

(c) Both ends fixed; $P_{cr} = \dfrac{4\pi^2 EI}{L^2}$

(d) One end fixed, the other pinned; $P_{cr} = \dfrac{(4.49)^2 EI}{L^2}$

Figure 3-3. Ideal columns with various boundary conditions.

3.3-4 Imperfect Columns

So far, we have concentrated on ideal columns. In practice, though, no column is of perfect geometry and the applied load does not necessarily pass through the centroid of the column cross section. It is therefore necessary to study the behavior of columns of imperfect geometries and of columns for which the load is applied eccentrically. Another reason for studying the behavior of columns of imperfect geometries is that by letting the imperfection disappear (limiting process), we can predict the behavior of the perfect system.

1. Eccentrically Loaded Columns. Consider first the case of a simply supported column (see Fig. 3-4) loaded eccentrically with the same eccentricity e at both ends. The equilibrium equation for this case is

$$w_{,xxxx} + k^2 w_{,xx} = 0 \tag{36a}$$

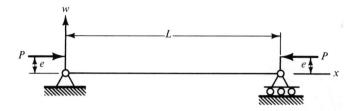

Figure 3-4. Eccentrically loaded column.

The boundary conditions are given by

$$\left.\begin{array}{l} w(0) = 0 \\ w(L) = 0 \\ EIw_{,xx}(0) = Pe \Rightarrow w_{,xx}(0) = k^2 e \\ EIw_{,xx}(L) = Pe \Rightarrow w_{,xx}(L) = k^2 e \end{array}\right\} \tag{36b}$$

The solution for this equation is

$$w = A_1 \sin k_x + A_2 \cos k_x + A_3 k + A_4 \tag{19}$$

Use of the boundary conditions, Eqs. (36b), yields

$$w(x) = e\left[1 - \cos kx - \tan \frac{kL}{2} \sin kx\right] \tag{37}$$

Note that as the load P increases from zero, the value of k, and consequently $\tan KL/2$, increases. Therefore the displacement function w becomes unbounded as $\tan KL/2$ approaches infinity and the corresponding load P approaches $\pi^2 EI/L^2$.

If Eq. (37) is evaluated at some characteristic point (say $x = L/2$), it may serve as a load-displacement relation. Denoting by δ the displacement at the midpoint, we may write

$$\delta = -e\left[1 - \cos \frac{kL}{2} - \tan \frac{kL}{2} \sin \frac{kL}{2}\right]$$

or

$$\delta = e\left[\sec \sqrt{\frac{P}{EI}} \frac{L}{2} - 1\right] \tag{38}$$

This result is plotted in Fig. 3-5, which shows that as $e \longrightarrow 0$, the plot represents the behavior of the perfect system.

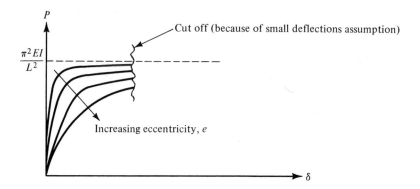

Figure 3-5. P-S diagram with eccentricity effect.

Finally, if the eccentricities are equal in magnitude but opposite in direction, the last two of the boundary conditions, Eqs. (36b), are given by

$$w_{,xx}(0) = -k^2 e \atop w_{,xx}(L) = k^2 e \Bigg\}$$

(39)

and the solution is

$$w(x) = e\left[-1 + \frac{2x}{L} + \cos kx - \cot \frac{kL}{2} \sin kx\right]$$

(40)

For this case the displacement becomes unbounded when the load P approaches $4\pi^2 EI/L^2$, and

$$w\left(\frac{L}{2}\right) = 0$$

and

$$w\left(\frac{L}{4}\right) = \frac{e}{2}\left[\sec\sqrt{\frac{P}{EI}}\frac{L}{4} - 1\right]$$

(41)

2. Columns with Geometric Imperfections. Next consider a simply supported column with an initial geometric imperfection, $w_0(x)$, loaded along the axis joining the end points (see Fig. 3-6).

The governing differential equation for this case becomes

$$EI(w_{,xx} - w_{0,xx}) + Pw = 0$$

(42)

Note that the change in curvature for this case is $(-w_{,xx} + w_{0,xx})$.

The boundary conditions for this problem are $w(0) = w(L) = 0$. Note that $w_0(0) = w_0(L) = 0$.

Since the initial shape is continuous and has a finite number of maxima

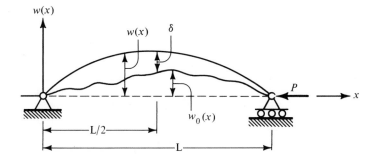

Figure 3-6. Imperfect column.

and minima in the range $0 < x < L$, then the shape can be represented by a sine series (there is no initial moment at the boundaries).

$$w_0(x) = \sum_{n=1}^{\infty} a_n \sin \frac{n\pi x}{L}$$

Then the differential equation, Eq. (42), assumes the following form

$$EI w_{,xx} + Pw = EI \sum_{n=1}^{\infty} \left(\frac{n\pi}{L}\right)^2 a_n \sin \frac{n\pi x}{L} \qquad (43)$$

A series solution may be assumed, each term of which satisfies the boundary conditions

$$w(x) = \sum_{j=1}^{\infty} A_j \sin \frac{j\pi x}{L} \qquad (44)$$

Substitution into Eq. (43) leads to the following equation:

$$-\left[EI\left(\frac{n\pi}{L}\right)^2 - P\right] A_n = EI\left(\frac{n\pi}{L}\right)^2 a_n \qquad (45)$$

Thus the solution becomes

$$w(x) = \sum_{n=1}^{\infty} \frac{P_n}{P_n - P} a_n \sin \frac{n\pi x}{L} \qquad (46)$$

and

$$w - w_0 = \sum_{n=1}^{\infty} \frac{P}{P_n - P} a_n \sin \frac{n\pi x}{L} \qquad (47)$$

where

$$P_n = \frac{n^2 \pi^2 EI}{L^2}$$

Southwell (Ref. 5) considered this problem of initial geometric imperfections and he concluded that, as long as the imperfection is such that a_1 exists, then a critical condition arises as $P \longrightarrow \pi^2 EI/L^2$. Furthermore, since in a test one can measure P and the net deflection of the midpoint $[w(L/2) - w_0(L/2) = \delta]$, he devised a plot from which $P_1 = P_E$ can be determined through use of the experimental data. This plot is known as the Southwell plot.

Assume that $a_1 \neq 0$ and all $a_n = 0$ for $n = 2, 3, 4, \ldots$. Then

$$\delta = w\left(\frac{L}{2}\right) - w_0\left(\frac{L}{2}\right) = \frac{Pa_1}{P_1 - P} \tag{48}$$

From this we obtain

$$P_1\left(\frac{\delta}{P}\right) - a_1 = \delta \tag{49}$$

And thus δ/P is a linear function of δ.

Since, as $P \longrightarrow P_1$, the first term of the series is the predominant one, Eq. (47), then for large values of P (but $P < P_1$) it can safely be assumed that Eq. (49) holds and δ/P varies linearly with δ. Thus, if for a test we plot δ/P vs δ at the higher values of δ, the relation is linear and the intercepts give a_1 and a_1/P_1 (see Fig. 3-7).

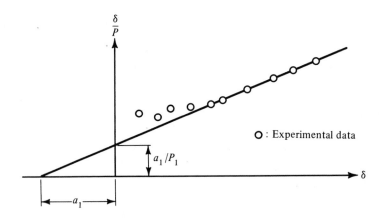

Figure 3-7. The Southwell plot.

3.3-5   Tilting of Forces

In many practical applications where stability of columns is a basic design criterion, the force does not remain fixed in direction, but it passes through a fixed point. This is a case of load-behavior (during the buckling process) problems where the system may still be considered as conservative. Note that the follower-force problem does not fall in this category. Examples of the types of problems are shown in Figs. 3-8 and 3-9.

Consider an elastic column, as shown in Fig. 3-10, with the applied load P passing through a fixed point C. Using the free end as the origin, for displacements and x-coordinate the governing equation (equilibrium) is

$$M = EIw_{,xx} = -(P \cos \varphi)w - (P \sin \varphi)x$$

If $\varphi =$ small constant (this implies that P is initially slightly tilted), the equation becomes

$$w_{,xx} + k^2 w = -k^2 \varphi x \qquad (50)$$

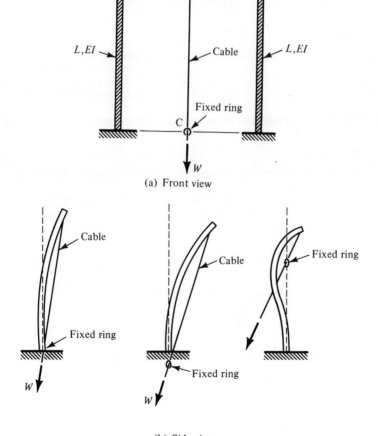

(a) Front view

(b) Side views

Figure 3-8. Loaded frame.

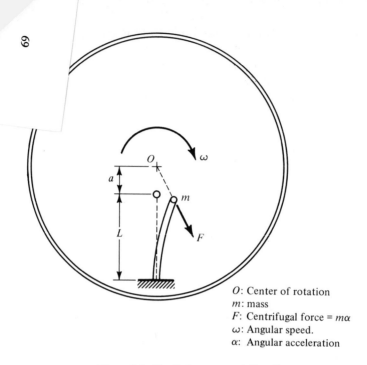

O: Center of rotation
m: mass
F: Centrifugal force = $m\alpha$
ω: Angular speed.
α: Angular acceleration

Figure 3-9. Elastic bar on a rotating disc.

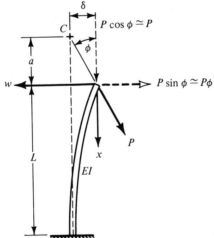

Figure 3-10. Tilt-buckled column.

The associated boundary conditions are

$$w(0) = 0, \qquad w_{,x}(L) = 0$$

and the auxiliary condition

$$w(L) = \delta \simeq a\varphi$$

The general solution of Eq. (50) is

$$w = -\varphi x + C_1 \sin kx + C_2 \cos kx \tag{51}$$

Use of the above conditions yields

$$a\varphi = -\varphi L \left[1 - \frac{\tan kL}{kL} \right] \tag{52}$$

Since we are only interested in the existence of a nontrivial solution, $\varphi \neq 0$, the characteristic equation is:

$$\tan kL = kL \left(1 + \frac{a}{L} \right) \tag{53}$$

This is a transcendental equation and it may be solved either numerically or graphically (see Fig. 3-11).

It is seen from Fig. 3-11 that the critical value for (kL) lies between zero and 4.493. Note the following special cases:

1. If $a = -L$, $(kL)_{cr} = \pi$ and $P_{cr} = \pi^2 EI/L^2$.
2. If $a = \pm\infty$, $(kL)_{cr} = \pi/2$ and $P_{cr} = \pi^2 EI/4L^2$.

An alternate solution is possible if we use the fourth-order differential equation, Eq. (18c). The solution to this equation is given by Eq. (19), or

$$w(x) = A_1 \sin kx + A_2 \cos kx + A_3 x + A_4 \tag{19}$$

Referring to Fig. 3-10 and placing the origin at the fixed end with x increasing toward the free end, we obtain the boundary conditions

$$w(0) = 0$$
$$w_{,x}(0) = 0$$
$$w_{,xx}(L) = 0$$
$$-[EIw_{,xxx}(L) + Pw_{,x}(L)] = -P\left[-\frac{w(L)}{a} \right]$$

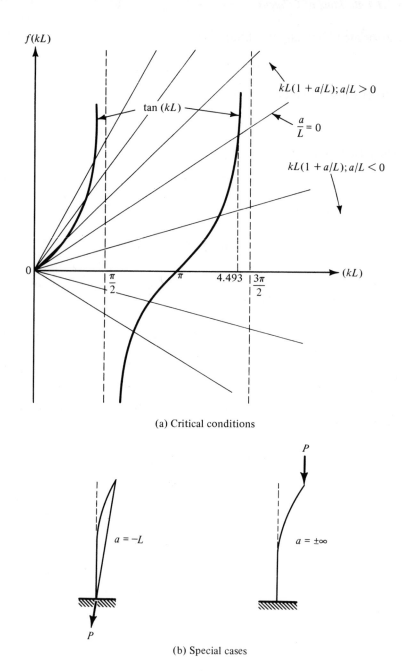

(a) Critical conditions

(b) Special cases

Figure 3-11. Critical conditions of tilt-buckling.

Use of the boundary conditions leads to the following characteristic equation

$$\begin{vmatrix} 0 & 1 & 0 & 1 \\ k & 0 & 1 & 0 \\ \sin kL & \cos kL & 0 & 0 \\ \sin kL & \cos kL & L & 1 \end{vmatrix} = 0$$

The expansion of this determinant is Eq. (53):

$$\tan kL = kL\left(1 + \frac{a}{L}\right) \qquad (53)$$

Tilt-buckling is also possible for other systems. Biezeno and Grammel (Ref. 6) discuss a number of cases of tilt-buckling in torsion and in bending. (See Fig. 3-12.)

Consider a shaft of torsional rigidity GJ, length L, fixed at one end, and carrying a rigid disc of diameter a, loaded as shown in Fig. 3-12a. When the disc is in the tilted position, the applied torque is $aP \sin \varphi$. The torque transmitted to the shaft is $(GJ)(\varphi/L)$. From equilibrium considerations

$$GJ\frac{\varphi}{L} = aP \sin \varphi \qquad (54)$$

It is clear that a bifurcation point exists at $\varphi = 0$ and $P = GJ/aL$ (see Fig. 3-12b). Thus

$$P_{cr} = \frac{GJ}{aL} \qquad (55)$$

Next, consider the beam shown in Fig. 3-12c. If $l_1 = L$, the governing differential equation becomes

$$EIw_{,xx} = -\frac{aP \sin \varphi}{L} x \qquad (56)$$

Two integrations with respect to x and the use of the boundary conditions $w(0) = w(L) = 0$ yield

$$w = \frac{aP}{6EIL}\varphi x(L^2 - x^2) \qquad (57)$$

Since $\varphi = -w_{,x}(L)$, then

$$\varphi = \frac{aPL}{3EI}\varphi \qquad (58)$$

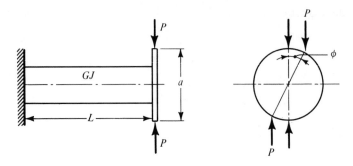

(a) Geometry of torsional shaft

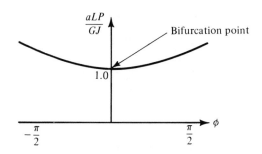

(b) Load-deflection curve for torsional tilt-buckling

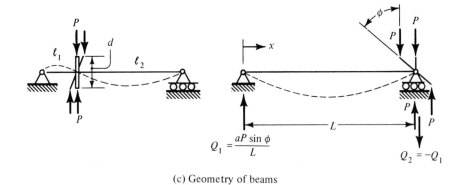

(c) Geometry of beams

Figure 3-12. Tilt-buckling of shafts and beams.

From this equation we see that a nontrivial solution exists if

$$P_{cr} = \frac{3EI}{aL} \tag{59a}$$

Simliarly, the critical load for the more general case is given by

$$P_{cr} = \frac{3EI}{a} \cdot \frac{l_1 + l_2}{l_1^2 - l_1 l_2 + l_2^2} \tag{59b}$$

Note that the result given by Eq. (59a) can be derived from Eq. (59b) if we let $l_1 = L$ and $l_2 = 0$.

3.3-6 Effect of Transverse Shear

One of the assumptions made in deriving the governing differential equations of beam bending (see Euler-Bernoulli assumptions) is that the effect of transverse shear on deformations is negligible. If this assumption is removed, the beam is called a Timoshenko beam, and the associated theory, the Timoshenko theory (see Refs. 7 and 8).

Consider a beam element dx units long which in the bent position rotates ψ units. If $w_{,x}$ denotes the rotation of the midsurface, the difference between ψ and $w_{,x}$ denotes the shear strain. (See Fig. 3-13.) Thus

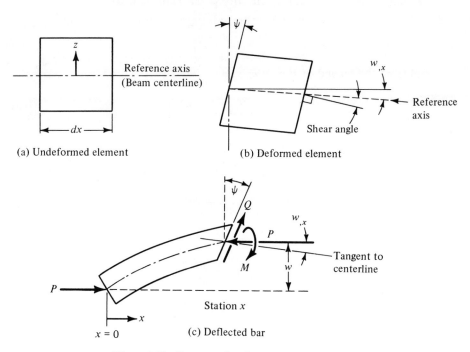

(a) Undeformed element   (b) Deformed element

(c) Deflected bar

Figure 3-13. Geometry for shear corrections.

$$\gamma = \psi - w,_x \tag{60}$$

The bending stress at any position z-units from the midsurface is given by

$$\sigma_x = Ez\psi,_x \tag{61}$$

Thus the governing differential equations of a simply supported column (in the bent position) are

$$\left. \begin{array}{l} EI\psi,_x + Pw = 0 \\ Q = -P\psi \end{array} \right\} \tag{62}$$

These two are the moment and force equilibrium equations. (See Fig. 3-13.)
If the shear strain were uniform, the shear force Q would be given by

$$Q = \tau A = GA(\psi - w,_x) \tag{63}$$

Since the strain is not uniform, then a correction factor, n, is used which depends on the cross-sectional geometry. For a solid rectangular cross section $n = 1.20$, for a solid circular cross section $n = 1.11$, and for an I-beam, $n = 1.2 \, (A/A_f)$, where A is the cross-sectional area and A_f is the area of the two flanges of the I-beam. If the correction factor is used

$$w,_x = \psi - \frac{nQ}{AG}$$

and from the second of Eqs. (62)

$$w,_x = \psi \left[1 + \frac{nP}{AG} \right] \tag{64}$$

Use of this equation and the first of Eqs. (62) leads to the following differential equation in a single dependent variable

$$\frac{EIw,_{xx}}{1 + nP/AG} + Pw = 0$$

or

$$w,_{xx} + \frac{P}{EI} \left[1 + \frac{nP}{AG} \right] w = 0 \tag{65}$$

Since the column is simply supported, the boundary conditions associated with the reduced-order equation, Eq. (65), are

$$w(0) = w(L) = 0 \tag{66}$$

It is easily concluded that the first eigenvalue is π and

$$P_{cr}\left(1 + \frac{nP_{cr}}{AG}\right) = \frac{\pi^2 EI}{L^2} = P_1 \tag{67}$$

Solving for P_{cr}, we obtain

$$P_{cr} = \frac{\sqrt{1 + 4nP_1/AG} - 1}{2n/AG} = \frac{2P_1}{1 + \sqrt{1 + 4nP_1/AG}} \tag{68}$$

Note that these two forms are equivalent.

Another expression is obtained for P_{cr} by a different approach (see Art. 2.17 of Ref. 8). This expression is

$$P_{cr} = \frac{P_1}{1 + P_1/AG} \tag{69}$$

where

$$P_1 = \frac{\pi^2 EI}{L^2}$$

3.4 THE KINETIC APPROACH

In using this approach, we are interested in the character of the motion (in the small) of the beam under constant P. In other words, if the column is compressed to any level of P, then the ends are made immovable (no additional axial deformation), and the column is given a small initial disturbance, what is the tendency of the system? If the system simply tends to oscillate about the undisturbed static equilibrium position, then the static equilibrium is stable. Before this can be accomplished, the differential equation governing the motion of a beam under the Euler-Bernoulli assumptions must be derived.

If m denotes the mass per unit length and if the effect of rotary inertia is neglected, then the equation of motion of the beam under constant axial load P is given by (see Refs. 7, 9, or 10)

$$(EIw_{,xx})_{,xx} + \bar{P}w_{,xx} + mw_{,tt} = 0 \tag{70}$$

If the ideal column is simply supported, then the solution of Eq. (70) is given by

$$w = f(t) \sin \frac{n\pi x}{L} \tag{71}$$

Note that the associated boundary conditions $w(0) = w(L) = w_{,xx}(0) = w_{,xx}(L) = 0$ are satisfied by this solution, Eq. (71).

Substitution into Eq. (77) yields

$$f_{,tt} + \left(\frac{n\pi}{L}\right)^2 \frac{1}{m}\left[\frac{n^2\pi^2EI}{L^2} - P\right]f = 0 \tag{72}$$

and if

$$\omega_n^2 = \frac{1}{m}\left(\frac{n\pi}{L}\right)^2\left[\frac{n^2\pi^2EI}{L^2} - P\right]$$

then

$$f_{,tt} + \omega_n^2 f = 0 \tag{73}$$

We see from Eq. (73) that, if $\omega_n^2 > 0$, then the motion is oscillatory, and if $\omega_n^2 < 0$, then the motion is diverging. Thus

$$P_{cr} = \frac{n^2\pi^2EI}{L^2} \qquad n = 1, 2, \ldots \tag{74}$$

and the smallest value corresponds to $n = 1$. Also note that the frequency of oscillations decreases as P approaches P_{cr}, while the frequency increases as the applied axial load increases in tension.

Another procedure that can be used here is as follows. Starting with Eq. (70), we may write the separated solution as

$$w(x, t) = g(x)e^{i\omega t} \tag{75}$$

where $\omega/2\pi$ is the frequency of small oscillations. Use of this solution leads to the following ordinary differential equation

or

$$\left.\begin{array}{c} EIg_{,xxxx} + Pg_{,xx} - m\omega^2 g = 0 \\[2mm] g_{,xxxx} + k^2 g_{,xx} - \dfrac{m\omega^2}{EI}g = 0 \end{array}\right\} \tag{76}$$

The general solution of this is

$$g(x) = \sum_{i=1}^{i=4} A_i e^{\lambda_i x} \tag{77}$$

where the A_i are constants and the λ_i are the roots of the biquadratic

$$\lambda^4 + k^2\lambda^2 - \frac{m\omega^2}{EI} = 0 \tag{78}$$

The λ_i roots are:

$$\left.\begin{array}{l} \lambda_{1,2} = \pm i\dfrac{k}{\sqrt{2}}\left[1 + \sqrt{1 + \dfrac{4m\omega^2}{k^4EI}}^{-1/2}\right] = \pm i\alpha \\[4mm] \lambda_{2,3} = \pm\dfrac{k}{\sqrt{2}}\left[-1 + \sqrt{1 + \dfrac{4m\omega^2}{k^4EI}}^{-1/2}\right] = \pm\beta \end{array}\right\} \tag{79}$$

The requirement of satisfying the prescribed boundary conditions leads to a set of four homogeneous algebraic equations in A_i. Since a nontrivial solution is sought, the determinant of the coefficient of the constants A_i must vanish. This is the characteristic equation.

To demonstrate this procedure, consider the case of a column with simply supported boundaries.

$$g(0) = g(L) = 0$$
$$g_{,xx}(0) = g_{,xx}(L) = 0 \tag{80}$$

The characteristic equation for this case becomes

$$\begin{vmatrix} A_1 & A_2 & A_3 & A_4 \\ 1 & 1 & 1 & 1 \\ -\alpha^2 & -\alpha^2 & \beta^2 & \beta^2 \\ e^{-i\alpha L} & e^{i\alpha L} & e^{-\beta L} & e^{\beta L} \\ -\alpha^2 e^{-i\alpha L} & -\alpha^2 e^{i\alpha L} & \beta^2 e^{\beta L} & \beta^2 e^{\beta L} \end{vmatrix} = 0 \tag{81}$$

Subtraction of the first column from the second, and the third from the fourth, and rearrangement of columns lead to the following

$$\begin{vmatrix} 1 & 1 & 0 & 0 \\ -\alpha^2 & -\beta^2 & 0 & 0 \\ e^{-i\alpha L} & e^{-\beta L} & e^{i\alpha L} - e^{-i\alpha L} & e^{\beta L} - e^{-\beta L} \\ -\alpha^2 e^{-i\alpha L} & \beta^2 e^{-\beta L} & -\alpha^2(e^{i\alpha L} - e^{-i\alpha L}) & \beta^2(e^{\beta L} - e^{-\beta L}) \end{vmatrix} = 0 \tag{82}$$

Expanding the above, we obtain

$$(\beta^2 + \alpha^2)^2 [e^{-i\alpha L} - e^{-i\alpha L}][e^{\beta L} - e^{-\beta L}] = 0 \tag{83}$$

and since

$$\left. \begin{aligned} e^{i\alpha L} - e^{-i\alpha L} &= 2i \sin \alpha L \\ e^{\beta L} - e^{-\beta L} &= 2 \sinh \beta L \end{aligned} \right\} \tag{84}$$

then Eq. (83) becomes

$$4(\alpha^2 + \beta^2)^2 \sin \alpha L \sinh \beta L = 0 \tag{85}$$

It is easily seen that

$$4(\alpha^2 + \beta^2)^2 \sinh \beta L \neq 0 \tag{86}$$

Thus

$$\sin \alpha L = 0 \tag{87}$$

The solution of Eq. (87) leads to

$$\alpha L = n\pi \tag{88}$$

Replacing the expression for α and squaring both sides, we obtain

$$1 + \sqrt{1 + \frac{4m\omega^2}{k^4 EI}} = \frac{2n^2\pi^2}{L^2 k^2}$$

or

$$1 + \frac{4m\omega^2}{k^4 EI} = \frac{4n^4\pi^4}{L^4 k^4} - \frac{4n^2\pi^2}{L^2 k^2} + 1 \tag{89}$$

from which

$$\omega^2 = \frac{EI}{m}\left(\frac{n\pi}{L}\right)^2 \left[\left(\frac{n\pi}{L}\right)^2 - k^2\right] \tag{90}$$

We see from Eq. (90) that the motion ceases to be oscillatory as $\omega \rightarrow 0$; consequently, the static equilibrium point corresponding to $k(P)$ ceases to be stable. Thus

$$P = \frac{n^2\pi^2 EI}{L^2}$$

and the smallest load corresponds to $n = 1$ or

$$P_{cr} = \frac{\pi^2 EI}{L^2}$$

3.5   ELASTICALLY SUPPORTED COLUMNS

In most structural configurations, columns are supported by other structural elements which provide elastic types of restraints at the ends of the columns. These restraints are similar to spring supports of the rotational (1b-inches per radian) as well as the extensional type (1bs per inch). In many cases, by knowing the structural configuration supporting the column, we can accurately estimate the intensity (spring constant) of the corresponding spring. In this section, the characteristic equation for a spring-supported column is derived and solutions are presented for a number of special cases.

In Fig. 3-14, the column of length L has a constant stiffness EI. It is spring-supported at both ends and is loaded axially by a compressive load P. The buckling equation is, as before,

$$w_{,xxxx} + k^2 w_{,xx} = 0 \qquad k^2 = \frac{P}{EI} \tag{91}$$

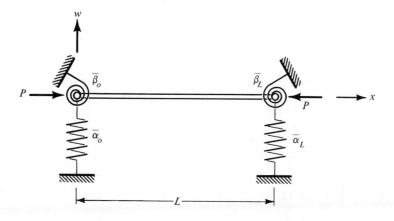

Figure 3-14. Elastically supported column.

and the boundary conditions are given by

$$\text{at } x = 0: \quad -[EIw_{,xxx} + Pw_{,x}] = \bar{\alpha}_0 w$$
$$EIw_{,xx} = \bar{\beta}_0 w_{,x} \tag{92a}$$

$$\text{at } x = L: \quad -[EIw_{,xxx} + Pw_{,x}] = -\bar{\alpha}_L w$$
$$EIw_{,xx} = -\bar{\beta}_L w_{,x} \tag{92b}$$

If we introduce the new parameters α_0, α_L, β_0, and β_L

$$\left. \begin{array}{cc} \alpha_0 = \dfrac{\bar{\alpha}_0}{EI} & \beta_0 = \dfrac{\bar{\beta}_0}{EI} \\[2mm] \alpha_L = \dfrac{\bar{\alpha}_L}{EI} & \beta_L = \dfrac{\bar{\beta}_L}{EI} \end{array} \right\} \tag{93}$$

the buckling equation and the associated boundary conditions become

$$w_{,xxxx} + k^2 w_{,xx} = 0 \tag{94}$$
$$w_{,xxx}(0) + k^2 w_{,x}(0) + \alpha_0 w(0) = 0 \tag{94a}$$
$$w_{,xx}(0) - \beta_0 w_{,x}(0) = 0 \tag{94b}$$
$$w_{,xxx}(L) + k^2 w_{,x}(L) - \alpha_L w(L) = 0 \tag{94c}$$
$$w_{,xx}(L) + \beta_L w_{,x}(L) = 0 \tag{94d}$$

Thus the problem has been reduced to an eigen-boundary-value problem.

The solution to the differential equation is given by Eq. (17), and the use of Eqs. (94) leads to the following four linear homogeneous algebraic equations in the A_i's ($i = 1, 2, 3, 4$):

$$\left.\begin{array}{l} \alpha_0 A_2 + k^2 A_3 + \alpha_0 A_4 = 0 \\ \beta_0 k A_1 + k^2 A_2 + \beta_0 A_3 = 0 \\ (\alpha_L \sin kL)A_1 + (\alpha_L \cos kL)A_2 - (k^2 - \alpha_L L)A_3 + \alpha_L A_4 = 0 \\ (k\beta_L \cos kL - k^2 \sin kL)A_1 - (\beta_L k \sin kL + k^2 \cos kL)A_2 + \beta_L A_3 = 0 \end{array}\right\} \quad (95)$$

The requirement of the existence of a nontrivial solution leads to the vanishing of the determinant

$$\begin{vmatrix} 0 & \alpha_1 & k^2 & \alpha_1 \\ \beta_0 k & k^2 & \beta_L & 0 \\ \alpha_L \sin kL & \alpha_L \cos kL & \alpha_L L - k^2 & \alpha_L \\ \begin{array}{l}(k\beta_L \cos kL \\ -k^2 \sin kL)\end{array} & \begin{array}{l}-(\beta_L k \sin kL \\ +k^2 \cos kL)\end{array} & \beta_L & 0 \end{vmatrix} = 0 \quad (96)$$

If we denote the quantity kL by u, the characteristic equation is given by the following transcendental equation:

$$\left[-(\alpha_0 + \alpha_L)\frac{u^6}{L^6} + \{\beta_0\beta_L(\alpha_0 + \alpha_L) + \alpha_0\alpha_L L\}\frac{u^4}{L^4} \right.$$

$$+ \left. \alpha_0\alpha_L(\beta_0 + \beta_L - \beta_0\beta_L L)\frac{u^2}{L^2} \right] \sin u$$

$$+ \left[(\alpha_0 + \alpha_L)(\beta_0 + \beta_L)\frac{u^5}{L^5} - \alpha_0\alpha_L L(\beta_0 + \beta_L)\frac{u^3}{L^3} \right.$$

$$- \left. 2\alpha_0\alpha_L\beta_0\beta_L\frac{u}{L} \right] \cos u + 2\alpha_0\alpha_L\beta_0\beta_L\frac{u}{L} = 0 \quad (97)$$

Solution of this transcendental equation (the smallest positive root) yields u_{cr}, from which we can compute P_{cr}:

$$P_{cr} = \frac{u_{cr}^2 EI}{L^2} \quad (98)$$

A number of special cases are reported below. These are (1) both ends pinned, (2) both ends clamped, (3) one end clamped and the other free, (4) one end clamped and the other pinned, and (5) both ends free against translation, constrained against rotation.

1. Both Ends Pinned. This case is characterized by $\alpha_0 = \alpha_L = \infty$ and $\beta_0 = \beta_L = 0$. If every term of Eq. (97) is divided by $\alpha_0\alpha_L$, and if we let $1/\alpha_0$, $1/\alpha_L$, β_0, and β_L approach zero, then

$$L\frac{u^4}{L^4} \sin u = 0 \quad (99)$$

From Eq. (99) $u_{cr} = \pi$ and

$$P_{cr} = \frac{\pi^2 EI}{L^2}$$

2. Both Ends Clamped. This case is characterized by $\alpha_0 = \alpha_L = \beta_0 = \beta_L = \infty$. If we divide Eq. (97) by $\alpha_0 \alpha_L \beta_0 \beta_L$ and take the limit as $1/\alpha_0$, $1/\alpha_L$, $1/\beta_0$, and $1/\beta_L$ approach zero, then

$$-L \frac{u^2}{L^2} \sin u - 2 \frac{u}{L} \cos u + 2 \frac{u}{L} = 0 \tag{100}$$

Since $u/L \neq 0$, then

$$\frac{u}{2} \sin u - (1 - \cos u) = 0$$

$$\frac{u}{2} 2 \sin \frac{u}{2} \cos \frac{u}{2} - 2 \sin^2 \frac{u}{2} = 0 \tag{101}$$

$$\left(\frac{u}{2} \cos \frac{u}{2} - \sin \frac{u}{2} \right) \sin \frac{u}{2} = 0$$

The smallest root is obtained from $\sin u/2 = 0$, or $u_{cr} = 2\pi$, from which

$$P_{cr} = \frac{4\pi^2 EI}{L^2}$$

3. One End Clamped, the Other Free. This special case is characterized by either $\alpha_0 = \beta_0 = 0$ and $\alpha_L = \beta_L = \infty$, or $\alpha_0 = \beta_0 = \infty$ and $\alpha_L = \beta_L = 0$. For either case, the characteristic equation reduces to

$$\frac{u^5}{L^5} \cos u = 0$$

from which $u_{cr} = \pi/2$ and $P_{cr} = \pi^2 EI/4L^2$.

4. One End Clamped, the Other Pinned. This case is characterized by either

$$\alpha_0 = \beta_0 = \alpha_L = \infty \quad \text{and} \quad \beta_L = 0 \quad \text{or} \quad \alpha_0 = \alpha_L = \beta_L = \infty \quad \text{and} \quad \beta_0 = 0$$

The characteristic equation reduces to

$$\left(\frac{u}{L} \right)^2 \sin u - L \left(\frac{u}{L} \right)^3 \cos u = 0 \tag{102}$$

Since $(u/L) \neq 0$, then

$$\tan u = u \tag{103}$$

The smallest root of Eq. (103) leads to the critical condition or

$$u_{cr} = 4.493 \quad \text{and} \quad P_{cr} = 20.2\frac{EI}{L^2}$$

5. Both Ends Free Against Translation, Constrained Against Rotation. This last special case is characterized by $\alpha_0 = \alpha_L = 0$ and $\beta_0 = \beta_L = \infty$. First divide Eq. (79) by $\beta_0\beta_L(\alpha_0 + \alpha_L)$. Then, since $1/\beta_0 = 1/\beta_L = 0$ and

$$\lim_{\substack{\alpha_0 \to 0 \\ \alpha_L \to 0}} \frac{\alpha_0\alpha_L}{\alpha_0 + \alpha_L} = 0$$

the characteristic equation reduces to

$$\frac{u^4}{L^4}\sin u = 0 \tag{104}$$

from which $u_{cr} = \pi$ and $P_{cr} = \pi^2 EI/L^2$.

3.6 CRITICAL SPRING STIFFNESS

To clearly demonstrate the meaning of critical spring stiffness, consider the following example. (See Fig. 3-15.) The column is pinned at the left end and supported with an extensional spring at the loaded right end. From the discussion of the previous section, this case is characterized by $\alpha_0 = \infty$, $\beta_0 = \beta_L = 0$, and α_L. The characteristic equation becomes

$$\left(\frac{u}{L}\right)^4\left[-\left(\frac{u}{L}\right)^2 + \alpha_L L\right]\sin u = 0 \tag{105}$$

Since $u/L \neq 0$, then

$$(-k^2 + \alpha_L L)\sin kL = 0$$

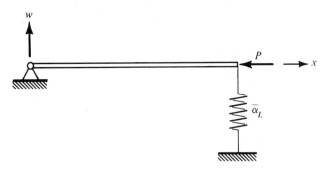

Figure 3-15. Critical spring stiffness model.

Thus either

$$P_{cr} = \bar{\alpha}_L L$$

or

$$P_{cr} = \frac{\pi^2 EI}{L^2} = P_E$$

Notice that for $\bar{\alpha}_L$ very small, $P_{cr} = \bar{\alpha}_L L < P_E$, but as $\bar{\alpha}_L$ increases, P_{cr} increases until $P_{cr} = \bar{\alpha}_L L = P_E$. This can happen when $\alpha_L = \pi^2 EI/L^3$. Any further increase in $\bar{\alpha}_L$ will yield $\bar{\alpha}_L L > P_E$, which implies that $P_{cr} = P_E$. This means that for $\bar{\alpha}_L > \pi^2 EI/L^3$, the column will always buckle in an Euler mode, and therefore there is no need to make the spring any stiffer than $\bar{\alpha}_L = \pi^2 EI/L^3$ because no increase in the critical load can result from it. Then the value $\pi^2 EI/L^3$ is called a critical spring stiffness.

Another case where there exists a critical spring stiffness is shown in Fig. 3-16. Consider the spring to act at the middle of the bar and the bar to deflect in a symmetric mode. From symmetry, the vertical reactions are $Q/2 = \alpha\delta/2$ (Fig. 3-16) and the reduced-order equation for the range $0 < x < L/2$ is

$$EIw_{,xx} + Pw = \frac{Q}{2} x \qquad (106)$$

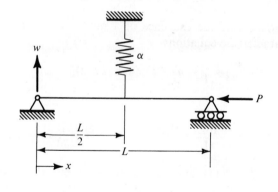

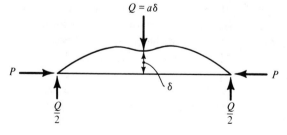

Figure 3-16. Elastically supported columns (spring at the midpoint).

The associated boundary and auxiliary conditions are

$$
\left.
\begin{aligned}
w(0) &= 0 \\
w_{,x}\left(\frac{L}{2}\right) &= 0 \\
w\left(\frac{L}{2}\right) &= \delta
\end{aligned}
\right\}
\tag{107}
$$

The solution is

$$
w = \frac{Q}{2P}\left(x - \frac{1}{k}\frac{\sin kx}{\cos kL/2}\right)
\tag{108}
$$

where

$$
k^2 = \frac{P}{EI}
$$

or

$$
w = \frac{\alpha\delta}{2P}\left(x - \frac{1}{k}\frac{\sin kx}{\cos kL/2}\right)
\tag{109}
$$

Using the auxiliary condition, we obtain

$$
\delta = \frac{\alpha\delta}{2Pk}\left[\frac{kL}{2} - \tan\frac{kL}{2}\right]
\tag{110}
$$

The requirement of the existence of a nontrivial solution leads to the following characteristic equation:

$$
1 = \frac{\alpha L}{4P}\left[1 - \frac{\tan(kL/2)}{kL/2}\right]
$$

or

$$
-\frac{16EI}{\alpha L^3}\left(\frac{kL}{2}\right)^2 = -1 + \frac{\tan(kL/2)}{kL/2}
\tag{111}
$$

We see from Eq. (111) that $(kL/2)_{cr}$ and consequently P_{cr} may be calculated for any given value of α. This may be done either numerically or graphically (see Fig. 3-17). Note from Fig. 3-15 that as $\alpha \to 0$, $(kL/2) \to \pi/2$, and $P_{cr} \to \pi^2 EI/L^2$ as expected. Furthermore, as $\alpha \to \infty$, $(kL/2)_{cr} \to 4.493$ and $P_{cr} \to (20.19)4EI/L^2$. However, if the bar were to buckle in an antisymmetric mode (with respect to $L/2$), then $P_{cr} = 4\pi^2 EI/L^2$; therefore there is no need for the spring to be any stiffer than $16\pi^2 EI/L^3$ (corresponding to $kL/2 = \pi$). This maximum value of spring stiffness, α, required for the bar to carry the maximum possible axial load ($= 4\pi^2 EI/L^2$) is called critical spring stiffness, or

$$
\alpha_{cr} = \frac{16\pi^2 EI}{L^3}
\tag{112}
$$

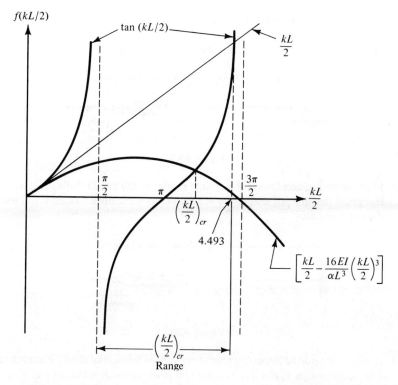

Figure 3-17. Critical conditions for spring-supported columns (at the midpoint).

If the elastic support is applied at l_1 units from the left end of the bar, then the characteristic equation is

$$-\frac{\sin kl_1 \sin kl_2}{Pk \sin kL} + \frac{l_1 l_2}{L} - \frac{1}{\alpha} = 0 \qquad (113)$$

where

$$l_2 = L - l_1$$

For details, see art. 2.6 of Ref. 8.

PROBLEMS

1. Calculate the critical load for an ideal column of length l and the following boundary conditions:
   (a) $w(0) = w(l) = 0$
       $w_{,xx}(0) = w_{,x}(l) = 0$
   (b) $w(0) = 0, \; w_{,x}(l) = 0$
       $w_{,xx}(0) = 0, \; w_{,xxx}(l) = 0$

2. An ideal column is pinned at one end and fixed to a rigid bar of length a at the other end. The second end of the rigid bar is pinned on rollers (see figure). Find the critical load and discuss the extreme cases ($a \rightarrow 0$ and $a \rightarrow \infty$).

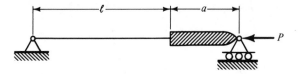

Figure P3-2.

3. Find the critical condition for a system similar to the one in Problem 2, with the exception that the left end of the ideal column is clamped (see figure). Discuss the extreme cases.

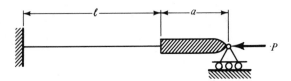

Figure P3-3.

4. A simply supported imperfect elastic bar carries horizontal thrust P at each end, with eccentricities e_1 and e_2. The initial line of centroids is curved and given by

$$w_0 = \sum_{n=1}^{\infty} a_n \sin \frac{n\pi x}{L}$$

Find the expression for the deflection w and the critical P value deduced from this.

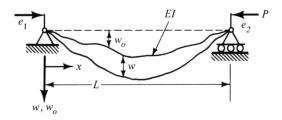

Figure P3-4.

5. An ideal column of length L is pinned at A and built in at B on a rigid disc of radius R, which is supported by an immovable frictionless pin at its center (see figure). Derive the characteristic equation and the expression for P_{cr}. Discuss the extreme cases ($R \rightarrow 0$ and $R \rightarrow \infty$).

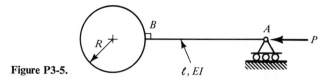

Figure P3-5.

6. Find an expression for the tilt-buckling load associated with the following two systems.

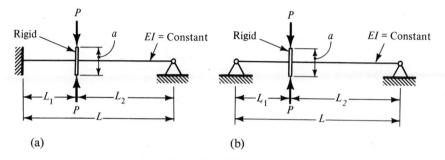

(a)   (b)

Figure P3-6.

7. A uniform disc of radius R rotates at constant angular velocity ω. A weightless elastic bar of length $L < R$ is fixed at one end and carries a mass m at the free end (see figure). Find the critical angular velocity at which buckling will occur.

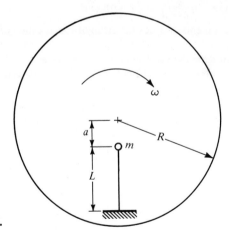

Figure P3-7.

8. A column is loaded by a tensile force T at a small angle φ to the vertical (see part a of the figure). Show by deriving and solving the differential equation that

$$\delta = l \tan \varphi \left(1 - \frac{\tan h\gamma l}{\gamma l}\right)$$

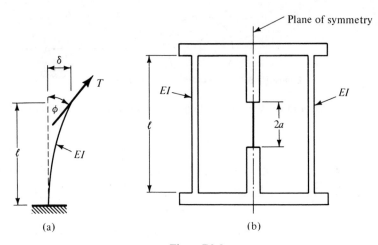

Figure P3-8.

where

$$\gamma^2 \simeq \frac{T}{EI}$$

The testing machine sketched (b of the figure) is applying compressive loading $2P$ to the test specimen. Show that buckling of the machine is possible, and indicate how the vertical load varies when a/l is made smaller. The machine columns are built in at both ends.

9. The top end of a flexible straight bar is attached by a stretched wire to a fixed point A. The initial tension T in the wire does not change appreciably when a small deflection δ occurs.

(a) Show that the critical values of P are determined by the equation

$$(kl)^2 = \frac{Tl^2}{EI}\left(\frac{l}{a} - \frac{\tan kl}{ka}\right)$$

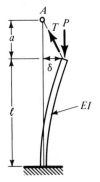

Figure P3-9.

where

$$k^2 \simeq \frac{P - T}{EI} \quad \text{and} \quad \sin \varphi \simeq \frac{\delta}{a}$$

(b) Check this result by taking $T = 0$.

10. The vertical bar AB is supported by an extensional spring of stiffness α at A. Explain what is meant by the critical value of such a spring stiffness, and find an expression for it when the lower end B of the bar is pinned. Suppose that the elastic bar is built in at B. Show that there is no critical spring stiffness for this case.

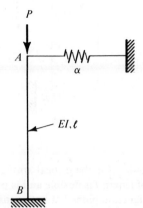

P

A   ⟋⟍⟋⟍⟋⟍
        α

← *EI,ℓ*

B

Figure P3-10.

11. A slender elastic column of length l is pinned at the left end, while the right end is restrained by a vertical support and a linear rotational spring of stiffness α. Derive the characteristic equation, and indicate on a sketch the range of the first root for all spring stiffnesses (from 0 to ∞).

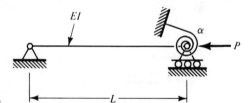

EI
            α
               ← *P*

Figure P3-11. |←   *L*   →|

12. A flexible uniform light blade AB is pinned at A to a base. The spiral represents a linear rotational spring. The blade carries a mass particle m at B. The base is now made to spin (at ω), in the plane, about a center O between A and B, and the blade is consequently under tension $mc\omega^2$.

(a) Examine the possibility of a critical speed ω at which a slight deflection of the blade becomes possible. Show that this possibility exists if the equation

$$\frac{EI}{\alpha l} + \frac{\coth \gamma l}{\gamma l} = \frac{l}{l} \frac{1}{1 - c \, (\gamma l)^2}, \qquad \gamma^2 \simeq \frac{mc\omega^2}{EI}$$

has a real root for γl.

(b) Consider the extreme case $\alpha \to \infty$ (clamped end). Prove that the "buck-ling" can occur when the center of rotation lies between A and B, but it can *not* occur when O is below A.

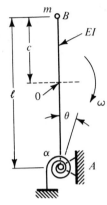

Figure P3-12.

13.

(a) Find an expression for the critical spring stiffness for the system shown. The member of length l is flexible and is pinned at both ends. The member of length a is rigid and pinned at both ends.

(b) The left end of the flexible bar, instead of being pinned, is fixed into a block that can move horizontally (frictionless rollers). Show that the characteristic equation for a given value α is

$$1 - \frac{\tan kl}{kl} = \frac{(kl)^2}{\alpha l^3 / EI - (l/a)(kl)^2}$$

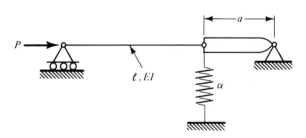

Figure P3-13.

14. By use of the kinetic approach, find the critical compressive force for an ideal column with the following boundary conditions:
   (α) Both ends fixed.
   (β) One end fixed, the other pinned.
   (γ) One end fixed, the other free.

REFERENCES

1. CARSLAW, H. S., and JAEGER, J. C., *Operational Methods in Applied Mathematics*, Dover Publications, Inc., New York, 1947, Chap. 11.

2. SHAMES, I. H., *Mechanics of Deformable Solids*, Prentice-Hall, Inc., Englewood Cliffs, N. J., 1964, Chap. 7.

3. HOFF, N. J., *The Analysis of Structures*, John Wiley & Sons, Inc., New York, 1956.

4. TIMOSHENKO, S. P., *History of Strength of Materials*, McGraw-Hill Book Co., New York, 1953, pp. 30–36.

5. SOUTHWELL, R. V., *An Introduction to the Theory of Elasticity*, Oxford at the Clarendon Press, 1936, p. 425.

6. BIEZENO, C. B., and GRAMMEL R., *Engineering Dynamics*, Vol. II, Part IV, Blackie and Son Ltd., London, 1956, p. 428.

7. THOMSON, WM. T., *Vibration Theory and Applications*, Prentice-Hall, Inc., Englewood Cliffs, N. J., 1965, p. 276.

8. TIMOSHENKO, S. P., and GERE, J. M., *Theory of Elastic Stability*, McGraw-Hill Book Co., New York, 1961, p. 135.

9. CHURCHILL, R. V., *Operational Mathematics*, McGraw-Hill Book Co., New York, 1958, p. 252.

10. COURANT, R., and HILBERT, D., *Methods of Mathematical Physics*, Vol. I, Interscience Publishers, Inc., New York, 1953, p. 295.

4

BUCKLING
OF FRAMES

Frames of various types, especially the civil-engineering type, are widely used in structural configurations such as buildings and bridges. These frames are subjected to concentrated and distributed loads which, in many cases, may cause buckling of an element or group of elements of the frame. Because the members are rigidly connected to other members, flexural deformations in one element cause deformations in the neighboring elements. This results in a loss of flexural rigidity of the entire system. Knowledge of the critical condition is essential in the design of both simple and complex frames.

This chapter is intended to familiarize the student with buckling of some simple frames, and it presents a few of the methods that can successfully be used to arrive at the critical condition. A more complete presentation of the buckling analysis of frames may be found in the books of Bleich (Ref. 1) and Britvec (Ref. 2). Since one of the methods employed for the analysis of frames is based on the theory of beam-columns, a review section will first be presented (see also Ref. 3).

4.1 BEAM-COLUMN THEORY

A slender bar meeting the Euler-Bernoulli assumptions under transverse loads as well as an in-plane compressive load (see Fig. 4-1) is called a *beam-*

95

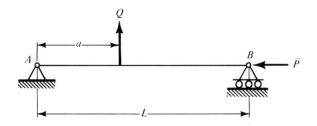

Figure 4-1. Beam-column.

column. The equation governing the response of a beam-column was derived in Chapter 3:

$$(EIw_{,xx})_{,xx} + \bar{P}w_{,xx} = q(x) + \sum_{i=1}^{n} Q_i\delta(x - x_i) + \sum_{j=1}^{m} C_j\eta(x - x_j) \quad (1)$$

The moment, M, and shear, V, at any station x are given by the following equations:

$$\begin{aligned} M &= EIw_{,xx} \\ V &= -[(EIw_{,xx})_{,x} + \bar{P}w_{,x}] \end{aligned} \quad (2)$$

The solutions to a number of problems are presented, and some of these solutions will be used in the buckling analysis of frames.

4.1-1 Beam-Column with a Concentrated Lateral Load

A simply supported beam-column under the application of a concentrated lateral load, Q, at station $x = a$ is shown in Fig. 4-1. The bending stiffness, EI, of the beam-column is taken to be constant.

The governing differential equation and the proper boundary conditions are given by

$$EIw_{,xxxx} + \bar{P}w_{,xx} = Q\delta(x - a) \quad (3)$$

$$\begin{aligned} w(0) &= w(L) = 0 \\ w_{,xx}(0) &= w_{,xx}(L) = 0 \end{aligned} \quad (4)$$

If we now separate the interval $[0, L]$ into two regions $0 < x < a$ and $a < x < L$, and if we denote by $w^1(x)$ and $w^2(x)$ the displacements in the two intervals, respectively, the differential equations and proper boundary conditions are given by

$$EIw^1_{,xxxx} + \bar{P}w^1_{,xx} = 0 \quad (5)$$

$$EIw^2_{,xxxx} + \bar{P}w^2_{,xx} = 0 \quad (6)$$

$$\begin{aligned} w^1(0) &= 0 \qquad w^1_{,xx}(0) = 0 \\ w^2(L) &= 0 \qquad w^2_{,xx}(L) = 0 \end{aligned} \quad (7)$$

The solutions to Eqs. (5) and (6) are

$$w^1(x) = A_1 \sin kx + A_2 \cos ky + A_3 x + A_4 \tag{8}$$
$$w^2(x) = B_1 \sin kx + B_2 \cos kx + B_3 x + B_4 \tag{9}$$

where $k^2 = \bar{P}/EI$.

There are eight constants to be evaluated, A_i and B_i $(i = 1, 2, 3, 4)$. These constants may be evaluated by use of the boundary conditions, Eqs. (7), and the auxiliary conditions at $x = a$. The auxiliary conditions are based on the fact that, at $x = a$, the deflection, slope, and moment must be continuous and the shear is discontinuous by a known amount $(\Delta V = Q)$.

The auxiliary conditions are

$$w^1(a) = w^2(a)$$
$$w^1_{,x}(a) = w^2_{,x}(a)$$
$$w^1_{,xx}(a) = w^2_{,xx}(a) \tag{10}$$
$$-[EIw^1_{,xxx}(a) + \bar{P}w^1_{,x}(a)] = -[EIw^2_{,xxx}(a) + \bar{P}w^2_{,x}(a)] + Q$$

Use of the eight equations, Eqs. (7) and (10), leads to the following solution:

$$w^1(x) = \frac{Q \sin k(L - a)}{\bar{P}k \sin kL} \sin kx - \frac{Q}{\bar{P}}\left(1 - \frac{a}{L}\right)x \qquad 0 \le x \le a$$
$$w^2(x) = \frac{Q \sin ka}{\bar{P}k \sin kL} \sin k(L - x) - \frac{Qa}{\bar{P}}\left(1 - \frac{x}{L}\right) \qquad a \le x \le L \tag{11}$$

By differentiation of Eq. (11), we obtain the following expressions for the slope and curvature (approximate):

$$w^1_{,x} = \frac{Q \sin k(L - a)}{\bar{P} \sin kL} \cos kx - \frac{Q(L - a)}{\bar{P}L} \qquad 0 \le x \le a \tag{12}$$

$$w^2_{,x} = -\frac{Q \sin ka}{\bar{P} \sin kL} \cos k(L - x) + \frac{Qa}{\bar{P}L} \qquad a \le x \le L \tag{13}$$

$$w^1_{,xx} = -\frac{Qk \sin k(L - a)}{\bar{P} \sin kL} \sin kx \qquad 0 \le x \le a \tag{14}$$

$$w^2_{,xx} = -\frac{Qk \sin ka}{\bar{P} \sin kL} \sin k(L - x) \qquad a \le x \le L \tag{15}$$

In the particular case for which $a = L/2$, the expressions for the deflection become

$$w^1 = \frac{Q \sin (kL/2)}{\bar{P}k \sin kL} \sin kx - \frac{Q}{2\bar{P}} x \tag{16a}$$

$$w^2 = \frac{Q \sin (kL/2)}{\bar{P}k \sin kL} \sin k(L - x) - \frac{QL}{2\bar{P}}\left(1 - \frac{x}{L}\right) \qquad (16b)$$

The maximum deflection occurs at $x = L/2$, and the expression for it is

$$w\left(\frac{L}{2}\right) = \delta = \frac{Q}{2\bar{P}k}\left(\tan \frac{kL}{2} - \frac{kL}{2}\right) \qquad (17)$$

In the absence of the in-plane load $\bar{P}$, the expression for the maximum deflection (at $L/2$) is

$$\delta_0 = \frac{QL^3}{48EI} \qquad (18)$$

If we rearrange the terms in Eq. (17) and make use of Eq. (18), the expression for the maximum deflection of the beam-column becomes

$$\delta = \delta_0 \frac{3(\tan u - u)}{u^3} \qquad (19)$$

where $u = kL/2$. Introducing the following notation

$$\chi(u) = 3\frac{\tan u - u}{u^3} \qquad (20)$$

the expression for δ becomes

$$\delta = \delta_0\chi(u) \qquad (21)$$

Numerical values of $\chi(u)$ are found in Appendix A of Ref. 1 for different values of u, and $\chi(u)$ is plotted versus u in Fig. 4-2 for $0 \leq u < \pi/2$. The factor $\chi(u)$ in Eq. (21) gives the influence of the in-plane load on the maximum deflection for the beam-column. From Fig. 4-2, we can decide up to what values of the axial thrust the neglect of its effect becomes unacceptable when we are interested in finding the deflection of this beam-column.

For this particular case ($a = L/2$), the rotation at $x = 0$ is given by

$$\frac{dw}{dx}\bigg|_{x=0} = \theta_A = \frac{QL^2}{16EI}\lambda(u) \qquad (22)$$

where

$$\lambda(u) = \frac{2(1 - \cos u)}{u^2 \cos u} \qquad (23)$$

Here again, the first factor in Eq. (22) denotes the slope at $x = 0$ in the absence of the in-plane load $\bar{P}$, and $\lambda(u)$ gives the influence of in-plane load $\bar{P}$.

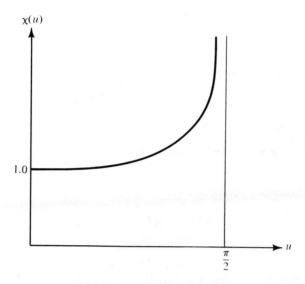

Figure 4-2. A plot of $\chi(u)$ versus u.

Numerical values of $\lambda(u)$ may be found in Appendix A of Ref. 1. The plot of $\lambda(u)$ versus u resembles that of $\chi(u)$ versus u (see Fig. 4-2).

Finally, the expression for the bending moment at $x = L/2$ for this particular case is

$$M_{\max} = EI\left(\frac{d^2w}{dx^2}\right)_{x=L/2} = -\frac{QL}{4}\frac{\tan u}{u} \tag{24}$$

Note that as $\bar{P}$ approaches zero, u approaches zero, and since

$$\lim_{u\to 0}\frac{\tan u}{u} = 1 \tag{25}$$

the expression for the maximum bending moment for the beam is

$$M_{0_{\max}} = -\frac{QL}{4}$$

4.1-2 Beam-Column with Two End-Couples

Consider the simply supported beam-column shown in Fig. 4-3 and loaded by two end-couples, M_A and M_B. The differential equation governing equilibrium is

$$EI\frac{d^4w}{dx^4} + \bar{P}\frac{d^2w}{dx^2} = 0 \tag{26}$$

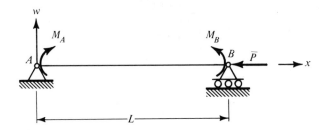

Figure 4-3. Beam-column loaded by two end couples.

The proper boundary conditions are

$$w(0) = w(L) = 0$$
$$EIw_{,xx}(0) = M_A \tag{27}$$
$$EIw_{,xx}(L) = M_B$$

The solution to Eq. (26) is given by

$$w(x) = A_1 \sin kx + A_2 \cos kx + A_3 x + A_4 \tag{28}$$

Use of the boundary conditions leads to the following expression for $w(x)$:

$$w(x) = \frac{M_A}{P}\left[\frac{L-x}{L} - \frac{\sin k(L-x)}{\sin kL}\right] + \frac{M_B}{P}\left[\frac{x}{L} - \frac{\sin kx}{\sin kL}\right] \tag{29}$$

Denoting by θ_A and θ_B the magnitudes of the rotation angles at A and B, respectively, we obtain

$$\theta_A = -\frac{dw}{dx}\bigg|_{x=0} = \frac{M_A L}{3EI}\psi(u) + \frac{M_B L}{6EI}\varphi(u) \tag{30}$$

$$\theta_B = \frac{dw}{dx}\bigg|_{x=L} = \frac{M_B L}{3EI}\psi(u) + \frac{M_A L}{6EI}\varphi(u) \tag{31}$$

where

$$\varphi(u) = \frac{3}{u}\left[\frac{1}{\sin 2u} - \frac{1}{2u}\right]$$

$$\psi(u) = \frac{3}{2u}\left[\frac{1}{2u} - \frac{1}{\tan 2u}\right] \tag{32}$$

$$u = \frac{kl}{2}$$

As before, the factors $\psi(u)$ and $\varphi(u)$ give the influence of the in-plane load on the end rotations. This means that in the absence of the in-plane load $\bar{P}$, the end rotations θ_{A_0} and θ_{B_0} are given by

$$\theta_{A_0} = \frac{M_A L}{3EI} + \frac{M_B L}{6EI}$$
$$\theta_{B_0} = \frac{M_B L}{3EI} + \frac{M_A L}{6EI} \tag{33}$$

4.1-3 Superposition

Since beam-column problems are nonlinear problems, superposition of solutions does not hold in the usual manner. The results can be superimposed if and only if the axial load $\bar{P}$ is the same for two or more cases of different lateral loads. To demonstrate the point, consider a simply supported beam-column of length L loaded first by a lateral loading $q_1(x)$ and second by $q_2(x)$. Let the response of the system to the two loadings be denoted by $w^1(x)$ and $w^2(x)$, respectively. Then the equilibrium equations for the two problems are:

$$EIw^1_{,xxxx} + \bar{P}w^1_{,xx} = q_1(x) \tag{34}$$
$$EIw^2_{,xxxx} + \bar{P}w^2_{,xx} = q_2(x) \tag{35}$$

By addition, we obtain

$$EI(w^1 + w^2)_{,xxxx} + \bar{P}(w^1 + w^2)_{,xx} = q_1(x) + q_2(x) \tag{36}$$

Next, consider the case of simultaneous application of the loadings $q_1(x)$ and $q_2(x)$. For this case the equilibrium equation is:

$$EIw_{,xxxx} + \bar{P}w_{,xx} = q_1(x) + q_2(x) \tag{37}$$

By comparison of Eqs. (36) and (37), it is clear that superposition holds for this type of problem. Thus superposition holds for any number of transverse loadings (distributed and concentrated forces, and applied moments) provided the in-plane force is the same and the beam-column is supported in the same manner for all loading cases (see Fig. 4-4).

4.2 APPLICATION OF BEAM-COLUMN THEORY TO THE BUCKLING OF ROTATIONALLY RESTRAINED COLUMNS

Consider a column which is supported against transverse translation at both ends but is restrained against rotation through rotational springs (see Fig. 4-5a). The problem here is to find $\bar{P}_{cr}$ as a function of the structural geometry (EI, L, β_0, and β_L).

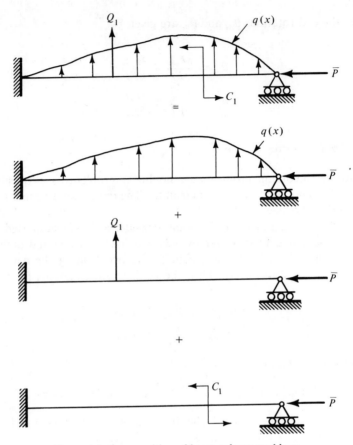

Figure 4-4. Superposition of beam-column problems.

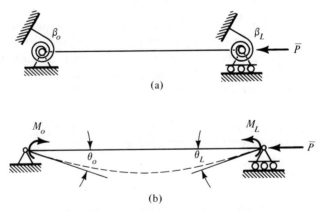

Figure 4-5. Rotationally restrained column.

Instead of this problem, we may consider the beam-column problem of Fig. 4-5b. According to the results of Section 4.1.2, the end rotations are given by

$$\theta_0 = \frac{M_0 L}{3EI}\psi(u) + \frac{M_L L}{6EI}\varphi(u) \tag{38}$$

$$\theta_L = \frac{M_0 L}{6EI}\varphi(u) + \frac{M_L L}{3EI}\psi(u) \tag{39}$$

By comparison of the beam-column problem to the original one, we may write

$$\begin{aligned} M_0 &= -\beta_0 \theta_0 \\ M_L &= -\beta_L \theta_L \end{aligned} \tag{40}$$

Substitution into Eqs. (38) and (39) yields

$$M_0\left[\frac{1}{\beta_0} + \frac{L}{3EI}\psi(u)\right] + M_L\left[\frac{L}{6EI}\varphi(u)\right] = 0$$
$$M_0\left[\frac{L}{6EI}\varphi(u)\right] + M_L\left[\frac{1}{\beta_L} + \frac{L}{3EI}\psi(u)\right] = 0 \tag{41}$$

These are two homogeneous linear algebraic equations in M_0 and M_L. For a nontrivial solution to exist (bifurcation), the determinant of the coefficients must vanish. Thus the characteristic equation is:

$$\left[\frac{1}{\beta_0} + \frac{L\psi(u)}{3EI}\right]\left[\frac{1}{\beta_L} + \frac{L\psi(u)}{3EI}\right] - \left[\frac{L\varphi(u)}{6EI}\right]^2 = 0 \tag{42}$$

where $u = kL/2$ and $k^2 = \bar{P}/EI$.

In the special case where $\beta_0 = \beta_L = \beta$, Eq. (42) becomes

$$\frac{1}{\beta} + \frac{1\psi(u)}{3EI} \pm \frac{L\varphi(u)}{6EI} = 0 \tag{43}$$

From the first of Eqs. (41), it is seen that

$$M_L = -M_0\left[\frac{1}{\beta} + \frac{L\psi(u)}{3EI}\right]\left[\frac{L\varphi(u)}{6EI}\right]^{-1} \tag{44}$$

Therefore we see that the plus sign in Eq. (43) corresponds to the symmetric case ($M_0 = M_L$), and the minus sign to the antisymmetric case ($M_0 = -M_L$).

For the symmetric case, substitution for the expressions $\psi(u)$ and $\varphi(u)$ leads to the following characteristic equation:

$$\tan u = -\frac{2EI}{\beta L}u \tag{45}$$

Figure 4-6a shows that u_{cr}, depending on the value of β, lies between $\pi/2$ and π. When $\beta \to 0$, $u_{cr} \to \pi/2$ and $P_{cr} = \pi^2 EI/L^2$ (both ends simply supported). When $\beta \to \infty$, $u_{cr} \to \pi$ and $P_{cr} = 4\pi^2 EI/L^2$ (both ends clamped).

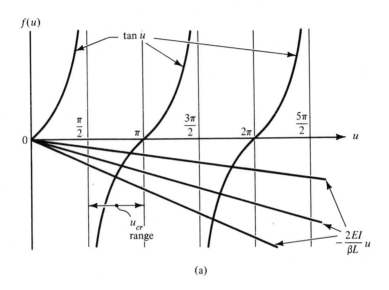

(a)

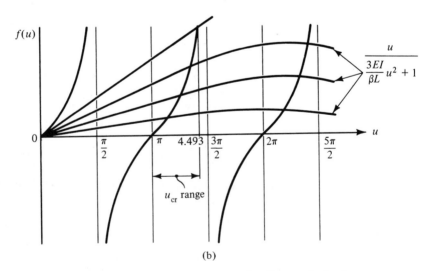

(b)

Figure 4-6. Critical conditions for rotationally restrained columns.

For the antisymmetric case, substitution for the expressions $\psi(u)$ and $\varphi(u)$ yields

$$\tan u = \frac{u}{\left[1 + \dfrac{2EI}{\beta L}u^2\right]} \tag{46}$$

From Fig. 4-6b we see that u lies between π and 4.493. When $\beta \to 0$, $u_{cr} \to \pi$ and $P_{cr} = 4\pi^2 EI/L^2$ (both ends simply supported). When $\beta \to \infty$, $u_{cr} \to$ 4.493 and $P_{cr} = 4(4.493)^2 EI/L^2$ (both ends clamped).

Therefore, for this special case $(\beta_0 = \beta_L = \beta)$, the column will buckle in a symmetric mode $(\pi/2 < u_{cr} < \pi)$.

4.3 RECTANGULAR RIGID FRAMES

Consider the frame shown in Fig. 4-7 and note that, as load P increases quasistatically from zero, it is possible to reach some PQ combination for which the frame will buckle (bifurcation). It is also clear that buckling may be caused by the existence of only P or Q and that the mode of failure in any case can be either symmetric or antisymmetric (see Figs. 4-8a and 4-8b).

In this particular problem, each member is elastically restrained against rotation at the ends because of the rigid connection to the adjacent member. Therefore the method described in the previous section may be applied,

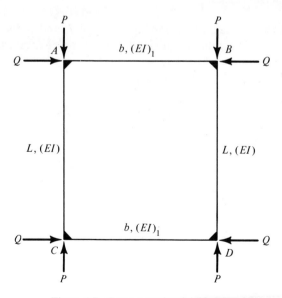

Figure 4-7. Geometry of a rigid frame.

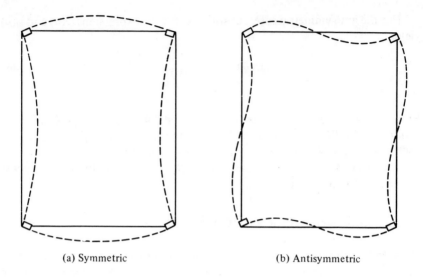

(a) Symmetric   (b) Antisymmetric

Figure 4-8. Buckling modes of a rigid frame.

provided the rotational spring constant can be expressed in terms of the structural geometry of the adjacent members.

Symmetric and antisymmetric buckling are treated separately in the following sections.

4.3-1   Symmetric Buckling

If we decompose the frame shown in Fig. 4-7, and if the frame is assumed to buckle in a symmetric mode, then the bending moments at the four corners are all equal (see Fig. 4-9).

Noting that $\theta_A = \theta_B$, from Eq. (38)

$$\theta_A = -\frac{Mb}{3(EI)_1}\psi(u_1) - \frac{Mb}{6(EI)_1}\varphi(u_1) \tag{47}$$

where

$$u_1 = k_1\frac{b}{2} \quad \text{and} \quad k_1^2 = \frac{Q}{(EI)_1}$$

From Eq. (47) we can obtain

$$\theta_A = -\frac{Mb}{2(EI)_1} \cdot \frac{\tan u_1}{u_1} \tag{48}$$

From Eq. (40), we obtain the expression for β. This equation is applicable because of the directions of moments and rotations as applied to AC (see Figs. 4-9 and 4-5b).

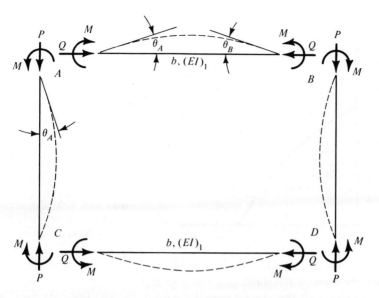

Figure 4-9. Symmetric buckling of the rigid frame.

$$\beta^{-1} = -\frac{\theta_A}{M} = \frac{b}{2(EI)_1} \frac{\tan u_1}{u_1} \tag{49}$$

Use of this expression for β in Eq. (43) for the vertical member (AC) yields (symmetric buckling)

$$\frac{b}{2(EI)_1} \frac{\tan u_1}{u_1} + \frac{L}{3EI}\psi(u) + \frac{L}{6EI}\varphi(u) = 0 \tag{50}$$

where

$$u = \frac{kL}{2} \quad \text{and} \quad k^2 = \frac{P}{EI}$$

Thus the characteristic equation becomes

$$\frac{L}{2EI} \frac{\tan u}{u} + \frac{b}{2(EI)_1} \frac{\tan u_1}{u_1} = 0 \tag{51}$$

or

$$\frac{\tan u}{u} = -\frac{EI}{(EI)_1} \frac{b}{L} \frac{\tan u_1}{u_1}$$

In the special case for which $(EI)_1/b = EI/L$, then

$$\frac{\tan u}{u} = -\frac{\tan u_1}{u_1} \tag{52}$$

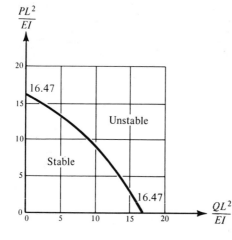

Figure 4-10. Critical conditions for a rigid square frame of constant stiffness.

The solution to this equation is plotted in Fig. 4-10 and it represents the boundary between the stable and unstable regions.

Consider also the special case for which $Q = 0$. For this case, since

$$\lim_{u_1 \to 0} \frac{\tan u_1}{u_1} = 1$$

the characteristic equation becomes

$$\frac{\tan u}{u} = -\frac{EIb}{(EI)_1 L} \tag{53}$$

Furthermore, if $EI = (EI)_1$ and $b = L$, then

$$\tan u = -u \tag{54}$$

The smallest root of this equation is 2.029; therefore

$$u_{cr} = \left(\frac{kL}{2}\right)_{cr} = 2.029$$

and   (55)

$$P_{cr} = 16.47 \frac{EI}{L^2}$$

4.3-2 Antisymmetric Buckling

If the frame buckles in an antisymmetric mode, then (see Fig. 4-11)

$$\theta_A = -\frac{Mb}{3(EI)_1} \psi(u_1) + \frac{Mb}{6(EI)_1} \varphi(u_1) \tag{56}$$

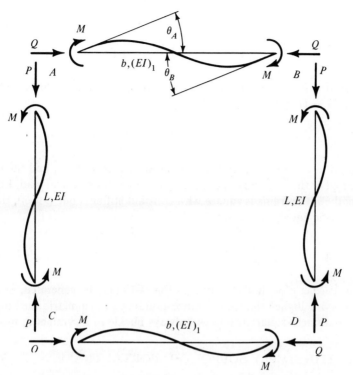

Figure 4-11. Antisymmetric buckling of the rigid frame.

From Eq. (56) and Eq. (40), we obtain

$$\frac{1}{\beta} = \frac{b}{6(EI)_1} \cdot \frac{3}{u_1}\left(\frac{1}{u_1} - \cot u_1\right) \tag{57}$$

Substitution of Eq. (57) into Eq. (43) yields the characteristic equation for antisymmetric buckling:

$$\frac{b}{(EI)_1} \frac{1}{u_1}\left(\frac{1}{u_1} - \cot u_1\right) = -\frac{L}{EI} \frac{1}{u}\left(\frac{1}{u} - \cot u\right) \tag{58}$$

Let us next consider a few special cases. First consider the case of $Q = 0$. Recognizing that $\psi(0) = \varphi(0) = 1$, then

$$\frac{1}{\beta} = \frac{b}{6(EI)_1} \tag{59}$$

The characteristic equation for this case is obtained if we substitute the expression for β, Eq. (59), into Eq. (43):

$$\frac{1}{u}\left(\frac{1}{u} - \cot u\right) = -\frac{EIb}{3(EI)_1L} \tag{60}$$

Furthermore, if we assume that the frame is square and of constant stiffness $[L = b, EI = (EI)_1]$,, then

$$\frac{1}{u} - \cot u = -\frac{u}{3} \tag{61}$$

The solution of this transcendental equation yields $u_{cr} > \pi$, and the critical load is higher than the corresponding symmetric mode critical load, Eq. (55).

Finally, if the frame is square with constant stiffness but $Q \neq 0$, then the characteristic equation becomes

$$\frac{1}{u_1}\left(\frac{1}{u_1} - \cot u_1\right) = -\frac{1}{u}\left(\frac{1}{u} - \cot u\right) \tag{62}$$

For this case, a plot similar to that in Fig. 4-10 may be generated. Since the intercepts are higher than those corresponding to symmetric buckling, the constant-stiffness square frame will always buckle in a symmetric mode.

4.4   THE SIMPLY SUPPORTED PORTAL FRAME

Let us consider the portal frame shown in Fig. 4-12. We are interested in finding the smallest possible load (P_{cr}) which will cause the frame to buckle. To accomplish this, we must consider all possible modes of buckling, compute P_{cr} for each mode, and establish, through a comparison, P_{cr} and the

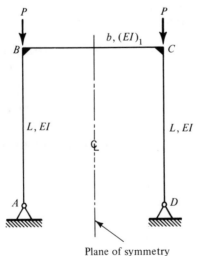

Plane of symmetry

Figure 4-12. Simply supported portal frame.

corresponding mode. Note that the frame is symmetric. The different buckling modes are shown in Fig. 4-13. Also note that there is no possibility of a sway buckling mode when the horizontal bar buckles symmetrically.

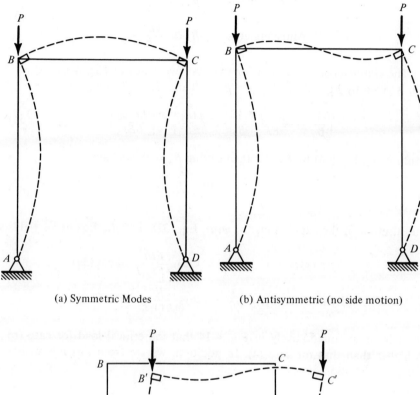

(a) Symmetric Modes   (b) Antisymmetric (no side motion)

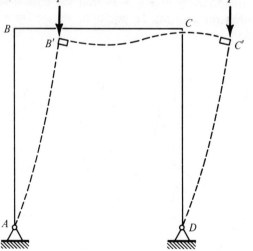

(c) Antisymmetric—sway buckling

Figure 4-13. Buckling modes for the simply supported portal frame.

First, the rotational elastic restraint provided to the vertical bars by the horizontal bar is the same as in the frame problem of Section 4.3 with $Q = 0$.

$$\text{Symmetric} \qquad \beta_B = \frac{2(EI)_1}{b}$$

$$\text{Antisymmetric} \qquad \beta_B = \frac{6(EI)_1}{b} \tag{63}$$

The characteristic equation for the first two cases of Fig. 4-13, (a) and (b), is given by Eq. (42) with $\beta_0 = \beta_A = 0$ and $\beta_L = \beta_B$, or

$$\left[\frac{1}{\beta_A} + \frac{L\psi(u)}{3EI}\right]\left[\frac{1}{\beta_B} + \frac{L\psi(u)}{3EI}\right] = \left[\frac{L\varphi(u)}{6EI}\right]^2 \tag{64}$$

Multiplying Eq. (64) by β_A and then setting $\beta_A = 0$, we have

$$\frac{1}{\beta_B} + \frac{L\psi(u)}{3EI} = 0 \tag{65}$$

Substitution of the expressions for $\psi(u)$, Eq. (32), and β_B, Eqs. (63), yields

$$\text{Symmetric (a)} \qquad \frac{1}{(2u)} + (2u)\frac{EIb}{2(EI)_1 L} = \cot(2u)$$

$$\text{Antisymmetric (b)} \qquad \frac{1}{2u} + (2u)\frac{EIb}{6(EI)_1 L} = \cot(2u) \tag{66}$$

It is shown qualitatively in Fig. 4-14 that the critical load for case (b) is higher than that for case (a). In addition, we see from this figure that $(2u)_{cr} > \pi$ for both cases.

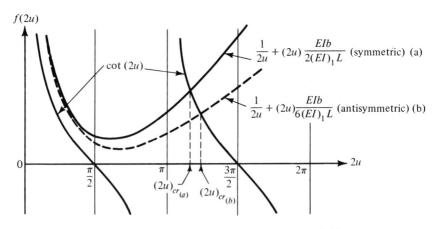

Figure 4-14. Critical conditions for cases (a) and (b).

As a special case of the characteristic equations, Eqs. (66), for this problem, we consider the horizontal bar to be extremely stiff. Then, $(EI)_1 \rightarrow \infty$ and the characteristic equation becomes

$$\frac{1}{2u} = \cot (2u) \quad \text{or} \quad \tan (2u) = (2u) \tag{67}$$

Since, $2u = kL$, the above results in

or
$$P_{cr} = 20.2 \frac{EI}{L^2}$$

This load represents the critical load for a column with one end fixed and the other simply supported.

The characteristic equation for the sway buckling case, Fig. 4-14c, cannot be obtained as a special case of Eq. (42) because point C is free to move in a direction normal to the column AB. Note that Eq. (42) represents the characteristic equation for a supported column with rotational end restraints, Fig. 4-5a.

The characteristic equation for the case of sway buckling may be obtained if we consider the column shown in Fig. 4-15. Note that the rotational restraint provided by the horizontal bar in Fig. 4-14c is $6(EI)_1/b$. The column of Fig. 4-15 is a special case of the elastically supported column treated in Chapter 3. Therefore, the characteristic equation for this model is obtained from Eq. (97) of Chapter 3 with the following expressions for the spring constants

$$\alpha_0 = \infty \qquad \alpha_L = 0 \qquad \beta_0 = 0 \qquad \beta_L = \frac{6(EI)_1/b}{EI}$$

In Eq. (97) of Chapter 3, the parameter u is defined by $u = kL$; therefore, wherever u appears, we must use $2u$.

Dividing Eq. (97) of Chapter 3 by α_0 and taking the limit as $1/\alpha_0 = \beta_0 = \alpha_L \rightarrow 0$, we have

$$-\frac{(2u)^6}{L^6} \sin (2u) + \frac{6EI_1}{bEI} \frac{(2u)^5}{L^5} \cos (2u) = 0$$

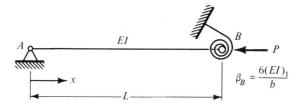

Figure 4-15. Model for sway buckling of simply supported portal frames.

And finally

$$(2u) \tan 2u = \frac{6(EI)_1 L}{EIb} \tag{68a}$$

or

$$\tan (2u) = \frac{6(EI)_1 L / EIb}{(2u)} \tag{68b}$$

From Fig. 4-16 we see that $2u_{cr} < \pi/2$, and the critical load for the simply supported portal frame is characterized by Eq. (68b). Therefore, as the load P is increased quasistatically from zero, the frame will sway buckle when P reaches the value that satisfies Eq. (68b).

Assuming that $EI_1 L = EIb$, we obtain

$$\tan 2u = \frac{3}{u} \tag{69}$$

from which

$$(2u)_{cr} = (kL)_{cr} = 1.35$$

and

$$P_{cr} = \frac{1.82 EI}{L^2}$$

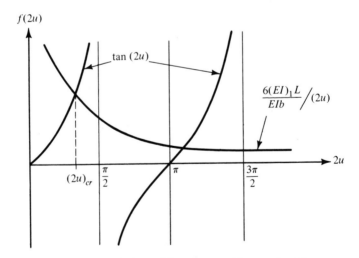

Figure 4-16. Critical conditions for case (c); sway buckling.

4.5   ALTERNATE APPROACH

We have already demonstrated in Section 4.4 that we may use Eq. (97) of Chapter 3 to obtain the characteristic equation for a frame by reducing the problem to an elastically restrained column. This approach may be used for any frame once the amount of elastic restraint has been determined either by beam theory or beam-column theory. This idea will be demonstrated through the rigid frame and then applied to some additional cases.

4.5-1 Rigid Frame

First consider the rigid frame shown in Fig. 4-7. First we reduce this problem to a column of length L, bending stiffness EI, and rotational restraints at the ends A and C of equal strength ($\bar{\beta}_A = \bar{\beta}_C$; see Fig. 4-17). Note that

$$\bar{\beta}_A = \bar{\beta}_C = \frac{2(EI)_1}{b} \cdot \frac{u_1}{\tan u_1}$$

from Eq. (49) for symmetric buckling (Fig. 4-9), and

$$\bar{\beta}_A = \bar{\beta}_C = \frac{[2(EI)_1/b]u_1}{(1/u_1) - \cot u_1}$$

from Eq. (57) for antisymmetric buckling (Fig. 4-11). To use Eq. (97) of Chapter 3, we must first recognize that wherever u appears in Eq. (97), we must use $2u$. Furthermore, the rotational restraint constants in Eq. (97) have been divided through by EI, or

$$\text{Symmetric} \qquad \beta_0 = \beta_L = \frac{\bar{\beta}_A}{EI} = \frac{2(EI)_1}{bEI} \frac{u_1}{\tan u_1}$$

$$\text{Antisymmetric} \qquad \beta_0 = \beta_L = \frac{\bar{\beta}_A}{EI} = \frac{[2(EI)_1/bEI]u_1}{(1/u_1) - \cot u_1} \qquad (70)$$

For the model shown in Fig. 4-17, we have $\alpha_0 = \alpha_L = \infty$, and $\beta_0 = \beta_L$ given by Eq. (70).

We first divide Eq. (97) by $\alpha_0\alpha_L$ and take the limit as $1/\alpha_0 = 1/\alpha_L = 0$. This leads to the following characteristic equation for the column model of Fig. 4-17.

$$\left[L\left(\frac{2u}{L}\right)^3 + (2\beta_0 - \beta_0^2 L)\left(\frac{2u}{L}\right)\right]\sin(2u)$$

$$- 2\beta_0\left[L\left(\frac{2u}{L}\right)^2 + \beta_0\right]\cos(2u) + 2\beta_0^2 = 0 \qquad (71)$$

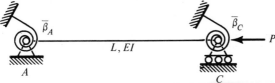

Figure 4-17. Column model for leg AC of the rigid frame.

If we first express the trigonometric functions in terms of the single angle, we have

$$\left[L\left(\frac{2u}{L}\right)^3 + (2\beta_0 - \beta_0^2 L)\left(\frac{2u}{L}\right)\right]2\sin u \cos u$$

$$- 2\beta_0 L\left(\frac{2u}{L}\right)^2(\cos^2 u - \sin^2 u) + 4\beta_0^2 \sin^2 u = 0 \qquad (72)$$

Next, if we divide Eq. (72) through by $\cos^2 u$, we obtain the following quadratic equation in $\tan u$:

$$\left[2\beta_0^2 + \beta_0 L\left(\frac{2u}{L}\right)^2\right]\tan^2 u + \left[L\left(\frac{2u}{L}\right)^3 + (2\beta_0 - \beta_0^2 L)\frac{2u}{L}\right]\tan u$$

$$- \beta_0 L\left(\frac{2u}{L}\right)^2 = 0 \qquad (73)$$

The solution for $\tan u$ by the quadratic formula is

$$\tan u = -\frac{2u}{L\beta_0} \qquad (74a)$$

$$\cot u = \frac{1}{u} + \frac{2u}{\beta_0 L} \qquad (74b)$$

It can be shown that Eq. (74a) corresponds to a symmetric mode, and therefore use of the corresponding expression for β_0 from Eqs. (70) yields

$$\frac{\tan u}{u} = -\frac{EIb}{(EI)_1 L} \cdot \frac{\tan u_1}{u_1} \qquad (75)$$

This is the same as Eq. (51), as expected.

Similarly, Eq. (74b) corresponds to the antisymmetric mode, and substitution for β_0 yields the following expression [see Eq. (57)]:

$$\frac{1}{u}\left(\frac{1}{u} - \cot u\right) = -\frac{2}{\beta_0 L} = -\frac{bEI}{(EI)_1 L}\frac{1}{u_1}\left(\frac{1}{u_1} - \cot u_1\right) \qquad (76)$$

4.5-2 The Clamped Portal Frame

Consider a portal frame similar to the one shown on Fig. 4-12 with the exception of having fixed supports at points A and D instead of simple supports. For this new problem, we must also consider all possible modes of failure. It is easily shown that the smallest load corresponds to sway buckling as demonstrated in Fig. 4-18. Therefore we will only find P_{cr} for sway buckling. The column model that characterizes this mode of failure is shown in Fig. 4-19.

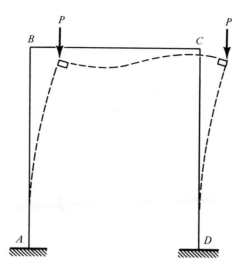

Figure 4-18. Sway-buckling of the clamped portal frame.

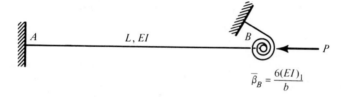

$$\bar{\beta}_B = \frac{6(EI)_1}{b}$$

Figure 4-19. Model for sway buckling of clamped portal frames.

The characteristic equation for this model is obtained from Eq. (97) of Chapter 3 with

$$\alpha_0 = \infty, \quad \beta_0 = \infty, \quad \alpha_L = 0 \quad \text{and} \quad \beta_L = \frac{6(EI)_1}{bEI}$$

Note again that we must divide Eq. (97) by $\alpha_0\beta_0$, take the limit as $1/\alpha_0 = 1/\beta_0 = \alpha_L = 0$, and use $2u$ instead of u. The characteristic equation is

$$\tan 2u = -\frac{2u}{\beta_L L} = -\frac{2ubEI}{6L(EI)_1} \tag{77}$$

From Fig. 4-20 we see that, depending on the value of $bEI/L(EI)_1$, the critical value for $(2u)$ varies between $\pi/2$ and π as expected.

If $bEI = (EI)_1 L$, then from Eq. (77)

$$\tan 2u = -\frac{2u}{6}$$

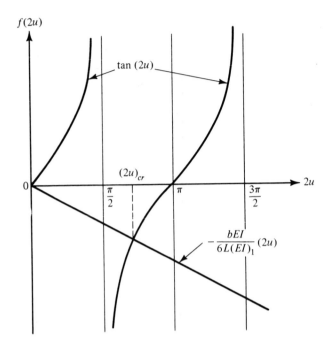

Figure 4-20. Critical conditions for sway buckling of clamped portal frames.

and

$$(2u)_{cr} = (kL)_{cr} = 2.71$$

from which

$$P_{cr} = \frac{7.344EI}{L^2}$$

4.5-3 Partial Frames

As a final application of the alternate approach, consider the partial frames of Fig. 4-21. The difference between these two partial frames is the support conditions at point A.

The elastic support provided by bar BC is a rotational spring with $\bar{\beta} = 4(EI)_1/b$, and an extensional spring, normal to the direction AB, with $\bar{\alpha} = (EA)_1/b$, where $(EA)_1$ is the extensional stiffness of bar BC. In most practical cases, $\bar{\alpha}$ is taken to be infinitely large. For such cases, the column models for the two partial frames are those shown in Fig. 4-22.

The characteristic equations for the two models are obtained from Eq. (97) of Chapter 3, with $\alpha_0 = \infty$, $\beta_0 = 0$, $\alpha_L = \infty$, $\beta_L = 4(EI)_1/bEI$ for case (*a*), and $\alpha_0 = \infty$, $\beta_0 = \infty$, $\alpha_L = \infty$, $\beta_L = 4(EI_1)/bEI$ for case (*b*). These

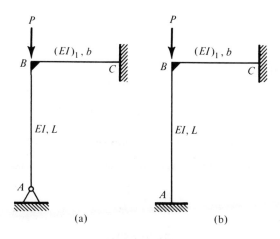

Figure 4-21. Partial frames.

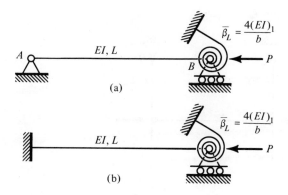

Figure 4-22. Column models for the partial frames.

equations are

case (a)

$$\left[L\left(\frac{2u}{L}\right)^4 + \beta_L\left(\frac{2u}{L}\right)^2\right]\sin 2u - L\beta_L\left(\frac{2u}{L}\right)^3\cos 2u = 0$$

or

$$\cot 2u = \frac{2u}{L\beta_L} + \frac{1}{2u}$$

$$\cot 2u = (2u)\frac{EIb}{4(EI)_1 L} + \frac{1}{2u} \tag{78}$$

case (b)

$$(1 - \beta_L L)\left(\frac{2u}{L}\right)^2\sin 2u$$

$$- \left[L\left(\frac{2u}{L}\right)^3 + 2\beta_L\left(\frac{2u}{L}\right)\right]\cos 2u + 2\beta_L\left(\frac{2u}{L}\right) = 0$$

or

$$(1 + \beta_L L)(2u) \sin 2u - L(2u)^2 \cos 2u + 2\beta_L L(1 - \cos 2u) = 0 \qquad (79)$$

where $\beta_L = 4(EI)_1/bEI$.

If $(EI)_1 = EI$ and $L = b$, the characteristic equations become

$$\text{case (a)} \qquad \cot 2u = \frac{2u}{4} + \frac{1}{2u} \qquad (80)$$

$$\text{case (b)} \qquad -3(2u) \sin 2u - (2u)^2 \cos 2u + 8(1 - \cos 2u) = 0 \qquad (81)$$

For this particular condition and for case (a), $(2u)_{cr} = (kl)_{cr} = 3.83$ and

$$P_{cr} = \frac{14.7EI}{L^2}$$

PROBLEMS

1. A horizontal column is rigidly attached to two vertical bars as shown. Find the expression for P_{cr} for buckling in the plane.

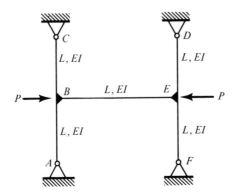

Figure P4-1.

2. The horizontal column BE is rigidly attached to two vertical bars AC and FD.

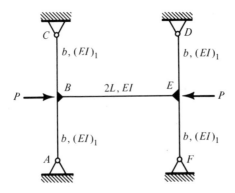

Figure P4-2.

The flexural rigidities EI and $(EI)_1$ are for deflections normal to the plane of the figure. If the value of $(EI)_1$ is low, the critical load for such normal deflections can be raised by increasing it. Show that such an improvement is possible only until $(EI)_1$ reaches the value $\pi^2 b^3 \, EI/24L^3$. The torsional rigidity of AC and FD is to be treated as negligibly small.

3. A horizontal column AB and a vertical bar CD are rigidly attached at C. Derive the characteristic equation and compute the value of $(L^2/EI)P_{cr}$ for the special case of $EI = EI_1$ and $2b = L$.

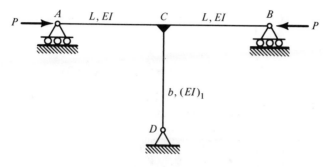

Figure P4-3.

4. Derive the characteristic equation for the cases shown in the figure.

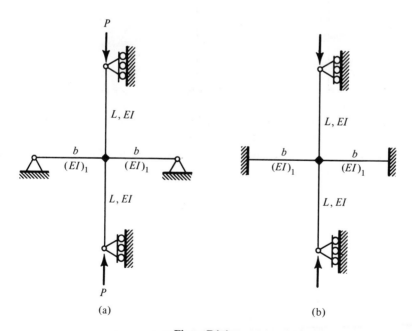

(a)   (b)

Figure P4-4.

5. Derive the characteristic equation for the rectangular rigid frame loaded as shown in the figure.

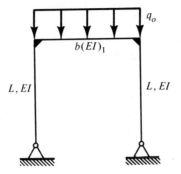

Figure P4-5.

6. Given a clamped portal frame loaded with a uniform loading q_0, derive the characteristic equation and compute q_{cr}.

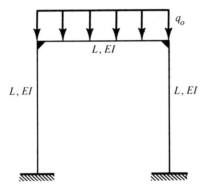

Figure P4-6.

7. Repeat Problem 6, replacing the uniform loading by a single concentrated load, P, at the plane of structural symmetry.

8. A vertical column AB is rigidly attached to a flexible horizontal bar CB. B is a

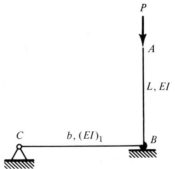

Figure P4-8.

roller that can turn without friction on a smooth base. The load P remains
vertical. Show that the characteristic equation is

$$kL \tan kL = \frac{3(EI)_1 L}{EIb}, \qquad k^2 = \frac{P}{EI}$$

Devise extreme cases to check this result.

9. Column AB and bar BC are identical and rigidly attached at B. The cross
   section is circular, with radius $R_0 (R_0 \ll L)$. Will the structure buckle in the
   plane of the figure or out of the plane?

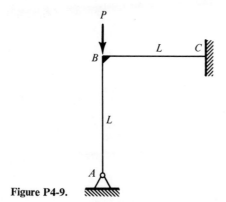

Figure P4-9.

10. Analyze the following partial frames for buckling in the plane of the figure.

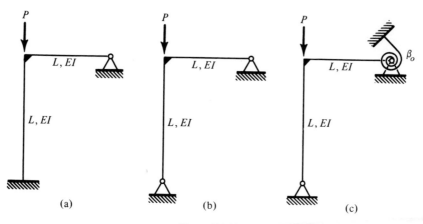

Figure P4-10.

REFERENCES

1. BLEICH, F., *Buckling Strength of Metal Structures*, edited by H. H. Bleich, McGraw-Hill Book Co., New York, 1952.

2. BRITVEC, S. J., *The Stability of Elastic Systems*, Pergamon Press, Elmsford, N.Y., 1973.

3. TIMOSHENKO, S. P., and GERE, J. M., *Theory of Elastic Stability*, McGraw-Hill Book Co., New York, 1961.

5

THE ENERGY CRITERION
AND ENERGY-BASED METHODS

5.1 REMARKS ON THE ENERGY CRITERION

As a basis for the energy criterion, we use the principle of the minimum total potential (see Appendix A). This principle states:

> Of all possible kinematically admissible deformation fields in an elastic conservative system, for a specified level of the external loads and the corresponding internal loads, only those that make the total potential assume a minimum value correspond to a stable equilibrium.

First of all, the system must be conservative for the principle to hold, which implies that the energy criterion holds only for such systems. The system is conservative if both the external and internal forces are conservative. Since we are dealing with an elastic system, the existence of a strain-energy density function (see Appendix A) implies that the internal forces are conservative. The external forces are conservative if the work done by these forces from state O to state I are independent of the path and depend only on the initial and final values of the kinematically admissible deformations (virtual displacements). The idea of a virtual displacement is discussed in detail in Appendix A.

This principle is an extension of the Lagrange-Dirichlet theorem to systems with infinitely many degrees of freedom. Although it is stated as a sufficiency condition for stable equilibrium, the energy criterion based upon this principle has been used as both a necessary and sufficient condition for stability.

In order to clearly state and apply the stability criterion, let $U_T[\bar{u}]$ be the total potential (functional) at an equilibrium position characterized by $\bar{u}$. Furthermore, let $U_T[\bar{u} + \epsilon \bar{u}_1]$ be the total potential in the neighborhood of the equilibrium position, where $\bar{u}_1$ denotes kinematically admissible deformations and ϵ is a small nonzero constant. If we now expand the integrals and group the integrals on the basis of powers of ϵ, then we may write

$$\Delta U_T = U_T[\bar{u} + \epsilon \bar{u}_1] - U_T[\bar{u}]$$
$$= \delta U_T[\bar{u}, \bar{u}_1, \epsilon] + \delta^2 U_T[\bar{u}, \bar{u}_1, \epsilon^2] + \cdots \qquad (1)$$

According to Eq. (1), δU_T, $\delta^2 U_T$, etc. denote the first, second, etc. variations in the total potential for kinematically admissible deformations.

For equilibrium, it is necessary that

$$\delta U_T = 0 \qquad (2)$$

for all $\bar{u}_1$. For a relative minimum (stable equilibrium), it is necessary that

$$\delta^2 U_T \geq 0 \qquad (3)$$

for all $\bar{u}_1$ (see Ref. 1). Note that, if the second variation is identically equal to zero for all $\bar{u}_i$, then no conclusion can be drawn and higher variations must be considered. For a relative minimum, since ϵ^3 can be positive or negative, the following two conditions must be satisfied

$$\delta^3 U_T \equiv 0$$
$$\delta^4 U_T \geq 0 \qquad (4)$$

for all $\bar{u}_i$. These steps are continued if $\delta^4 U_T \equiv 0$ for all $\bar{u}_i$ (see Ref. 2).

If we assume that the second variation is not identically equal to zero, then the stability criterion requires that $\delta^2 U_T$ be positive (definite) for every nonvanishing virtual displacement. Similarly, $\delta^2 U_T$ is negative definite for instability. The implication here is that $\delta^2 U_T$ changes its character at the critical load. Consequently, the critical load is the least value for which $\delta^2 U_T$ ceases to be positive definite and becomes positive semidefinite. This means that there exists at least one nonvanishing virtual displacement (buckling mode) for which the second variation is zero.

The energy criterion has been used by Timoshenko (Ref. 3) and Trefftz (Ref. 4) in various modified forms. The formulation suggested by Timoshenko is based on the following arguments: In a position of stable equilibrium, the total potential is a minimum, and consequently the increment in the total potential, ΔU_T, for every small kinematically admissible deformation from the equilibrium position, is positive. In terms of the strain energy and potential of the external forces,

$$\Delta U_i + \Delta U_p > 0 \tag{5}$$

Since $\Delta U_p = -\Delta W_e$, where ΔW_e is the work done by the external forces during these small deviations, Eq. (5) becomes

$$\Delta U_i > \Delta W_e \tag{6a}$$

This inequality becomes an equality as the load is increased to its critical value, and

$$\Delta U_i = \Delta W_e \tag{6b}$$

We can demonstrate this concept by applying it to the column problem. The Trefftz criterion, generalized herein, is based on the argument that, when the critical load is reached, $\delta^2 U_T$ becomes positive semidefinite. This means that the minimum value of $\delta^2 U_T$ becomes zero for certain nonvanishing virtual displacements. Now, if we set $\delta^2 U_T = V$, we want V to possess a minimum (which is zero) for certain nonvanishing virtual displacements (buckling mode). Thus, the first variation of V must be zero, or the first variation of the second variation of U_T must vanish. This criterion is applied to the column problem in this chapter and to the shallow arch in Chapter 7.

5.2 TIMOSHENKO'S METHOD

Timoshenko's method is fully outlined and applied to a number of systems in Arts. 2.8–2.10 of Ref. 3. This method, referred to by Timoshenko (Ref. 3) as the energy method, provides a shortcut to obtaining approximate but highly accurate values for the critical load. It avoids solving differential equations and becomes very useful when applied to systems with nonuniform stiffness, a case where the solution to the usual eigen-boundary-value problem is extremely difficult and in some cases impossible.

5.2-1 The Cantilever Column

Consider the cantilever shown in Fig. 5-1 under a constant directional thrust P applied quasistatically.

As the load is increased from zero, the work done by the force P is stored into the system as stretching strain energy. If we now consider that we allow a bending deformation, $w(x)$, which is very small so that it does not alter the stretching energy, the change in the total potential ΔU_T is given by

$$\Delta U_T = \Delta U_{i_B} + \Delta U_p \tag{7}$$

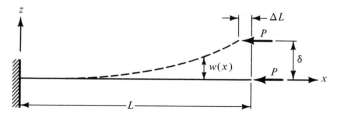

Figure 5-1. Geometry of the Cantilever Column.

where ΔU_{i_B} is the bending strain energy and ΔU_p is the change in the potential of the external force.

$$\Delta U_{i_B} = \tfrac{1}{2} \int_0^L EI(w'')^2 \, dx \qquad (8a)$$

$$\Delta U_p = -\Delta W_e = -P\Delta L = -\tfrac{1}{2}P \int_0^L (w')^2 \, dx \qquad (8b)$$

According to Timoshenko's argument, the straight configuration is stable if $\Delta U_T > 0$ and unstable if $\Delta U_T < 0$. A critical condition is reached when $\Delta U_T = 0$.

The additional steps are to assume a form for the admissible bending deformation, $w(x)$, and perform the indicated operations in the method. Let

$$w(x) = A\left(1 - \cos\frac{\pi x}{2L}\right) \qquad (9)$$

Note that A is an arbitrary constant and $1 - \cos(\pi x/2L)$ satisfies the kinematic boundary conditions at $x = 0$.

$$\Delta U_{i_B} = \frac{\pi^4 EI A^2}{64L^3}$$

and

$$\Delta U_p = -P\frac{\pi^2 A^2}{16L}$$

From this expression we obtain that

$$P_{cr} = \frac{\pi^2 EI}{4L^2}$$

which, of course, is the exact solution because the chosen deformation function happens to be the exact eigenfunction (see Chapter 3).

Next, let us use a different expression for $w(x)$:

$$w(x) = Ax^2 \qquad (10)$$

Then Timoshenko's method yields

$$P_{cr} = \frac{3EI}{L^2}$$

which is higher, by approximately 21.3%, than the exact value.

Finally, we use for $w(x)$ the shape corresponding to the solution of a cantilever loaded transversely by a concentrated load at the free end:

$$w(x) = Ax^2(3L - x) \tag{11}$$

Timoshenko's method then yields

$$P_{cr} = 2.5\frac{EI}{L^2}$$

This value is only 1.32% higher than the exact solution. The reason we get a better approximation in this case is because the expression for $w(x)$, given by Eq. (11), satisfies one of the natural boundary conditions at $x = L$, i.e., $M(L) = EIw''(L) = 0$. This condition is not satisfied by the expression for $w(x)$ given by Eq. (10).

5.2-2 The Simply Supported Column

In a similar manner, according to Timoshenko's method, a critical condition is reached, when

$$\frac{1}{2}\int_0^L EI(w'')^2\, dx = \frac{P}{2}\int_0^L (w')^2\, dx \tag{12}$$

Assume that $w(x)$ is given by a function that is kinematically admissible, i.e.,

$$w(x) = A(L - x)x \tag{13}$$

Then by Eq. (12)

$$P_{cr} = 12\frac{EI}{L^2}$$

This value is higher than the exact value by approximately 21.3%.

If we now choose a function that satisfies the natural boundary conditions as well, $M(0) = M(L) = 0$, we should expect the solution to improve.

Let

$$w(x) = A(L^3x - 2Lx^3 + x^4)$$
$$w'(x) = A(L^3 - 6Lx^2 + 4x^3)$$
$$w''(x) = A(12x^2 - 12Lx)$$

Then by Eq. (12)

$$P_{cr} = 9.88 \frac{EI}{L^2}$$

which is approximately 0.13 % higher than the exact value of the critical load.

5.2-3 The Rayleigh and Timoshenko Quotients

When Timoshenko (Ref. 1) applied his method to the cantilever column, he did not use, for the strain energy, the expression given by Eq. (7). Instead he made use of the fact that (see Fig. 5-1)

$$M(x) = P[\delta - w(x)] \tag{14}$$

and

$$(w'') = \frac{M}{EI}$$

Then

$$\Delta U_{i_B} = \frac{1}{2} \int_0^L \frac{M^2}{EI} \, dx \tag{15}$$

$$= \frac{P^2}{2} \int_0^L \frac{(\delta - w)^2}{EI} \, dx$$

According to his method, when P approaches P_{cr}

$$\frac{1}{2} P_{cr}^2 \int_0^L \frac{(\delta - w)^2}{EI} \, dx = \frac{P_{cr}}{2} \int_0^L (w')^2 \, dx \tag{16}$$

If we now use Eq. (10) for w, and employ Eq. (16) to solve for P_{cr}, we have

$$P_{cr} = 2.5 \frac{EI}{L^2}$$

which is much closer to the exact value than what we got before ($3EI/L^2$). Similarly, if the expression for w, given by Eq. (11), is used in Eq. (16)

$$P_{cr} = 2.4706 \frac{EI}{L^2}$$

This value, also, is closer to the exact value than what we got before. The improvement in the value, when using Eq. (16), occurs because the operation of differentiation indicated in Eq. (7) magnifies the error that exists when an approximate expression is used for $w(x)$.

We have seen so far that, as a consequence of Timoshenko's method, the expression for the critical load for any column may be written as

$$P = \frac{\int_0^L EI(w'')^2 \, dx}{\int_0^L (w')^2 \, dx} \tag{17}$$

This is called the integral or Rayleigh quotient, and it holds for all columns regardless of the boundary conditions. The name Rayleigh is used because a similar expression was derived and employed by Lord Rayleigh (Ref. 3) in the study of vibrations. Southwell (Ref. 4) outlines a procedure for finding buckling loads for a column which uses the Rayleigh quotient and refers to it as Rayleigh's method. Similar quotients may be derived for plates and certain shell configurations.

For the cantilever column, Eq. (16) is employed and the quotient becomes

$$P = \frac{\int_0^L (w')^2 \, dx}{\int_0^L \frac{(\delta - w)^2}{EI} \, dx} \tag{18}$$

Finally, if the column is simply supported, the reduced-order (second-degree) equation is applicable (see Chapter 3)

$$EIw'' + Pw = 0 \tag{19a}$$

and

$$w'' = -\frac{Pw}{EI} \tag{19b}$$

Substitution of this expression into the Rayleigh quotient yields

$$P = \frac{\int_0^L (w')^2 \, dx}{\int_0^L \frac{w^2}{EI} \, dx} \tag{20}$$

Quotients of the type given by Eqs. (18) and (20) are referred to as the Timoshenko quotient. Note that, when applicable, the Timoshenko quotient yields a closer approximation than the Rayleigh quotient.

Finally, when a column with elastic restraints at both ends is considered (see Fig. 3-14), the Rayleigh quotient becomes

$$P = \frac{\int_0^L EI(w'')^2 \, dx + \bar{\alpha}_0 w^2(0) + \bar{\alpha}_L w^2(L) + \bar{\beta}_0 w_{,x}^2(0) + \bar{\beta}_L w_{,x}^2(L)}{\int_0^L w_{,x}^2 \, dx} \tag{21}$$

Since the Timoshenko method leads to a quotient similar to that used by Lord Rayleigh, the method is often called the Rayleigh-Timoshenko method.

5.2-4   The General Rayleigh-Timoshenko Method

Starting with the concept of Timoshenko, we arrive at an integral quotient or the ratio of two functionals $I[u]$ and $J[u]$:

$$\lambda = \frac{I[u]}{J[u]} \tag{22}$$

We are interested in finding the function, u, which minimizes the quotient. If we use a series expression for u in the form of

$$u = \sum_{i=1}^{N} a_i g_i \tag{23}$$

where g_i are admissible functions and elements of a complete sequence (see Appendix A), the quotient becomes

$$\lambda = \frac{f_1(a_1, a_2, \ldots, a_n)}{f_2(a_1, a_2, \ldots, a_n)} \tag{24}$$

Now we must adjust the coefficients a_i in such a way that the ratio is a minimum. This requirement leads to

$$\frac{\partial \lambda}{\partial a_i} = \frac{1}{f_2^2}\left[\frac{\partial f_1}{\partial a_1}f_2 - \frac{\partial f_2}{\partial a_i}f_1\right] = 0 \tag{25}$$

Since f_2 is finite, use of Eq. (24) yields

$$\frac{\partial f_1}{\partial a_i} - \lambda\frac{\partial f_2}{\partial a_i} = 0, \qquad i = 1, \ldots, N \tag{26}$$

This procedure is referred to by many authors, including Timoshenko (Ref. 3), as the Ritz procedure.

As an example of this procedure, consider the cantilever of Fig. 5-1. If we employ the Rayleigh quotient,

$$I[w] = \int_0^L EI(w'')^2 \, dx$$

and $\tag{27}$

$$J[w] = \int_0^L (w')^2 \, dx$$

Let

$$w_A(x) = \sum_{n=1}^{3} a_n x^{n+1} \tag{28}$$

Then

$$f_1(a_1, a_2, a_3) = I[w_A]$$

$$= 4EI\left[a_1^2 L + 3a_1 a_2 L^2 + (3a_2^2 + 4a_1 a_3)L^3 + 9a_2 a_3 L^4 + \frac{36}{5} a_3^2 L^5 \right] \tag{29a}$$

and

$$f_2(a_1, a_2, a_3) = J[w_A] = \frac{4}{3} a_1^2 L^3 + 3a_1 a_2 L^4 + \left(\frac{9}{5} a_2^2 + \frac{16}{5} a_1 a_3 \right) L^5$$

$$+ 4a_2 a_3 L^6 + \frac{16}{7} a_3^2 L^7 \tag{29b}$$

If we define $\lambda = PL^2/EI$, Eqs. (26) become

$$8L(1 - \tfrac{1}{3}\lambda)a_1 + 3(4 - \lambda)L^2 a_2 + 16(1 - \tfrac{1}{5}\lambda)L^3 a_3 = 0$$
$$3(4 - \lambda)L^2 a_1 + 6(4 - \tfrac{3}{5}\lambda)L^3 a_2 + 4(9 - \lambda)L^4 a_3 = 0 \tag{30}$$
$$16(1 - \tfrac{1}{5}\lambda)L^3 a_1 + 4(9 - \lambda)L^4 a_2 + 32(\tfrac{9}{5} - \tfrac{1}{7}\lambda)L^5 a_3 = 0$$

Equations (30) represent a system of three linear homogeneous algebraic equations in a_1, a_2, and a_3. A nontrivial solution exists if the determinant of the coefficients vanishes.

$$\begin{vmatrix} 8(1 - \tfrac{1}{3}\lambda) & 3(4 - \lambda) & 4(1 - \tfrac{1}{5}\lambda) \\ 3(4 - \lambda) & 6(4 - \tfrac{3}{5}\lambda) & (9 - \lambda) \\ 4(1 - \tfrac{1}{5}\lambda) & (9 - \lambda) & 2(\tfrac{9}{5} - \tfrac{1}{7}\lambda) \end{vmatrix} = 0 \tag{31}$$

If only one term is considered ($a_1 \neq 0$, $a_2 = a_3 = 0$), then $\lambda_{cr} = 3$ and $P = 3EI/L^2$ as before. If two terms are considered ($a_1 \neq 0$, $a_2 \neq 0$, and $a_3 \equiv 0$), then $\lambda_{cr} = 2.48596$, which is only 0.75% higher than the exact solution. When all three terms are considered (a computer program was employed), $\lambda_{cr} = 2.4677$, which is extremely close to the correct answer (2.4674).

Note that every approximation is higher than the exact solution, and as more terms are considered, we converge to the minimum of λ from above. This, as expected, is true for all problems for which the formulation is characterized by Eq. (22). This means that the value of λ obtained by the use of an approximate expression for u in Eq. (22) cannot be any smaller than the value of λ corresponding to the exact expression for u.

5.2-5 The Nonuniform Stiffness Column

Consider a simply supported column with a bending stiffness given by (see Fig. 5-2)

$$EI = EI_0\left[1 + \frac{I_1}{I_0} \sin \frac{\pi x}{L}\right] \tag{32}$$

Furthermore, let the buckled deformation, $w(x)$, be approximated by

$$w_A(x) = A_1 \sin \frac{\alpha x}{L} + A_3 \sin \frac{3\pi x}{L} \tag{33}$$

If we use the general Rayleigh-Timoshenko procedure as outlined in the previous section, then

$$f_1(A_1, A_3) = EI_0\left(\frac{\pi}{L}\right)^4 \frac{L}{2}\left[A_1^2\left(1 + \frac{8}{3\pi}\frac{I_1}{I_0}\right) - A_1 A_3 18\left(\frac{8}{15\pi}\right)\frac{I_1}{I_0}\right.$$
$$\left. + A_3^2 81\left(1 + \frac{72}{35\pi}\frac{I_1}{I_0}\right)\right] \tag{34a}$$

$$f_2(A_1, A_3) = \left(\frac{\pi}{2}\right)^2 \frac{L}{2}[A_1^2 + 9A_3^2] \tag{34b}$$

Next, if we let $\lambda = P/P_{E_0}$, where $P_{E_0} = \pi^2 EI_0/L^2$, and employ Eqs. (26), we obtain the following system of equations in A_1 and A_3

$$\left.\begin{array}{l}\left[\left(1 + \frac{8}{3\pi}\frac{I_1}{I_0}\right) - \lambda\right]A_1 - 9\left(\frac{8}{15\pi}\right)\frac{I_1}{I_0}A_3 = 0 \\[2mm] -\frac{8}{15\pi}\frac{I_1}{I_0}A_1 + \left[9\left(1 + \frac{72}{35\pi}\frac{I_1}{I_0}\right) - \lambda\right]A_3 = 0\end{array}\right\} \tag{35}$$

For a nontrivial solution to exist, the determinant of the coefficients must

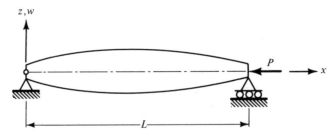

Figure 5-2. Simply supported column with nonuniform stiffness.

vanish. The expansion of the determinant yields the following quadratic equation in λ.

$$\lambda^2 - \lambda\left[10 + \frac{2224}{105\pi}\frac{I_1}{I_0}\right] + 9\left[1 + \frac{496}{105\pi}\frac{I_1}{I_0} + \frac{8192}{1575\pi^2}\frac{I_1^2}{I_0^2}\right] = 0 \quad (36)$$

From Eq. (36) we obtain λ_{cr}:

$$\lambda_{cr} = 5 + 3.371056\frac{I_1}{I_0} - 4\sqrt{1 + 1.261114\frac{I_1}{I_0} + 0.413813\frac{I_1^2}{I_0^2}} \quad (37)$$

Note that, if $w = A_1 \sin(\pi x/L)$ (one-term solution), then

$$\lambda_{cr} = \left(1 + \frac{8}{3\pi}\frac{I_1}{I_0}\right) \quad (38)$$

The actual values for λ are shown in Table 5-1 for various values of I_1/I_0. At this point there are two basic questions that we must answer. First, "Since the exact value for λ_{cr} is not known, does the two-term solution yield a good approximation for λ_{cr}?" Second, "Since there is considerable difficulty in obtaining λ_{cr} for a column with nonuniform flexural stiffness, why bother to build and analyze such columns?" Another way of stating the sec-

Table 5-1. Critical Loads for Nonuniform and Uniform Columns of Equal Weight

$\dfrac{I_1}{I_0}$	$\lambda_{NU_{cr}}$		$\lambda_{U_{cr}}$ $n=1$	$\lambda_{U_{cr}}$ $n=2$	$\lambda_{NU_{cr}}/\lambda_{U_{cr_{n=1}}}$		$\lambda_{NU_{cr}}/\lambda_{U_{cr_{n=2}}}$	
	One-Term Solution	Two-Term Solution			One-Term Solution	Two-Term Solution	One-Term Solution	Two-Term Solution
0	1.0000	1.0000	1.0000	1.0000	1.000	1.000	1.000	1.000
0.5	1.4244	1.4183	1.3183	1.3120	1.080	1.076	1.086	1.081
1.0	1.8488	1.8290	1.6366	1.6211	1.130	1.117	1.140+	1.128
2.0	2.6977	2.6405	2.2732	2.2265	1.187	1.162	1.212	1.186
4.0	4.3953	4.2488	3.5465	3.4193	1.239	1.198	1.285	1.243
6.0	6.0930	5.8505	4.8197	4.6013	1.264	1.214	1.324	1.271
8.0	7.7906	7.4497	6.0930	5.7781	1.279	1.223	1.348	1.289
10.0	9.4883	9.0478	7.3662	6.9518	1.288	1.228	1.365	1.302
20.0	17.9765	17.0320	13.7324	12.7996	1.309	1.240	1.404	1.331
30.0	26.4648	25.0130	20.0986	18.6331	1.317	1.244	1.420	1.342
40.0	34.9530	32.9932	26.4648	24.4610	1.321	1.247	1.429	1.349
50.0	43.4413	40.9730	32.8310	30.2860	1.323	1.248	1.434	1.353
100.0	85.8826	80.8703	64.6620	59.3925	1.328	1.251	1.446	1.362
∞					1.333	1.253	1.460	1.373

ond question is the following: "Is the critical load of a nonuniform stiffness column, $P_{NU_{cr}}$, higher than the critical load of a uniform stiffness column, $P_{U_{cr}}$, when the two columns are of equal weight?"

Before we answer these questions, we must consider the relation between the cross-sectional area, A, and the moment of inertia, I. If we let

$$I(x) = \alpha A^n(x) \tag{39}$$

where α is a constant, we note that, when $n = 1$, the width varies while the height remains constant; when $n = 2$, both the width and height vary in the same proportion; and when $n = 3$, the height varies while the width remains constant. For a column with a rectangular cross-section of height h and width b,

$$I = \frac{bh^3}{12} \quad \text{and} \quad A = bh$$

Now, for $n = 1$

$$\alpha = \frac{bh^3/12}{bh} = \frac{h^2}{12}$$

and for α to be a constant, h must be a constant.

Similarly, for $n = 2$,

$$\alpha = \frac{bh^3/12}{b^2h^2} = \frac{1}{12}\left(\frac{h}{b}\right)$$

and for α to be a constant, h/b must be a constant, which implies that the height and width vary proportionally.

Finally, for $n = 3$

$$\alpha = \frac{bh^3/12}{b^3h^3} = \frac{1}{12b^2}$$

and b must be a constant.

These conclusions are generally true for all symmetric cross-sections such as circular, elliptic, triangular, I-, and T-sections.

Returning to the two questions, we find the answer to the second question by comparing P_{cr} for the nonuniform geometry column with P_{cr} for the uniform column, provided the weights of the two are equal. However, since the two-term solution leads to a higher value for P_{cr} than the exact, this comparison is meaningful only if the two-term solution is a good approximation to the exact value of $P_{NU_{cr}}$. Let us first obtain the expressions for the volume (weight), $\bar{V}$, for the two columns.

For the uniform column

$$\bar{V} = \int_0^L A_U \, dx = A_U L \tag{40}$$

For the nonuniform column

$$\bar{V} = \int_0^L A_{NU}\, dx \tag{41}$$

Making use of Eqs. (32) and (39)

$$A_{NU} = \frac{1}{\alpha^{1/n}} I_0^{1/n} \left(1 + \frac{I_1}{I_0} \sin \frac{\pi x}{L}\right)^{1/n} \tag{42}$$

Substitution of Eq. (42) into Eq. (41) yields

$$\bar{V} = \left(\frac{I_0}{\alpha}\right)^{1/n} \int_0^L \left(1 + \frac{I_1}{I_0} \sin \frac{\pi x}{L}\right)^{1/n} dx \tag{43}$$

Let $\xi = \pi x/L$, then Eq. (43) becomes

$$\bar{V} = \left(\frac{I_0}{\alpha}\right)^{1/n} \frac{L}{\pi} \int_0^\pi \left(1 + \frac{I_1}{I_0} \sin \xi\right)^{1/n} d\xi \tag{44}$$

For a simply supported column $P_{U_{cr}}$ is given by the Euler load, or

$$P_{U_{cr}} = \frac{\pi^2 E I_U}{L^2}$$

Use of Eqs. (39) and (40) yields

$$P_{U_{cr}} = \frac{\pi^2 E}{L^2} \alpha \left(\frac{\bar{V}}{L}\right)^n \tag{45}$$

Through the use of Eq. (44), Eq. (45) becomes

$$P_{U_{cr}} = \frac{\pi^2 E I_0}{L^2} \left[\int_0^\pi \left(1 + \frac{I_1}{I_0} \sin \xi\right)^{1/n} \frac{d\xi}{\pi}\right]^n \tag{46}$$

From this equation

$$\lambda_{U_{cr}} = \left[\int_0^\pi \left(1 + \frac{I_1}{I_0} \sin \xi\right)^{1/n} \frac{d\xi}{\pi}\right]^n \tag{47}$$

For $n = 1$

$$\lambda_{U_{cr}} = \frac{1}{\pi} \int_0^\pi \left(1 + \frac{I_1}{I_0} \sin \xi\right) d\xi = 1 + \frac{2}{\pi} \frac{I_1}{I_0}$$

There are no closed-form solutions for $n = 2$ and 3. Values of $\lambda_{U_{cr}}$ are presented in Table 5-1 for $n = 1$ and $n = 2$ for a large range of I_1/I_0. In addition, the ratios of $\lambda_{NU_{cr}}$ to $\lambda_{U_{cr}}$ are presented for the one- and two-term solutions.

A number of investigators have dealt with the shape of the optimum column (see Refs. 6, 7, and 8). When there is no constraint on the stress level and the size of the area distribution, it is found that the optimum shape (for the simply supported column) starts with zero area at the ends and builds up to some maximum value at the center. Furthermore, the ratio of the critical load, corresponding to the optimum column, to the critical load for a uniform geometry column of equal volume is given by 1.216, 1.333, and 1.410 for $n = 1$, 2, and 3, respectively. A study of the results of Table 5-1 in connection with the above conclusions suggests that:

1. Since Eq. (32) does not necessarily correspond to the optimum shape (even with $I_0 = 0$), then $\lambda_{NU_{cr}}/\lambda_{U_{cr}}$ should be smaller than 1.216 (for $n = 1$) and 1.333 (for $n = 2$). Because it is not, the two-term solution has not converged to the exact value and more terms are needed in Eq. (33).

2. The more uniform the column is (smaller I_1/I_0 values) the better the convergence is.

Finally, we may conclusively state that nonuniformity in stiffness, of the type expressed by Eq. (32), yields a stronger configuration than that of a uniform geometry of the same weight.

5.3 THE RAYLEIGH-RITZ METHOD

The Rayleigh-Ritz or simply the Ritz method is explained in Appendix A. As far as buckling problems are concerned, there are two possible applications of the method.

The first type of application concerns problems for which a Rayleigh quotient exists (columns, plates, cylindrical shells, etc.). In this case, if the total potential, or some characteristic functional such as $\delta^2 U_T$ according to the Trefftz criterion, is expressed in the form of $U_T = I[u] - \lambda J[u]$, where λ denotes the eigenvalues (the lowest of which corresponds to the critical load parameter), the method suggests that we express u in terms of a series of the type

$$U_A = \sum_{i=1}^{N} a_i g_i$$

where g_i are kinematically admissible functions.
Then,

$$U_T = f_1(a_1, a_2, \ldots, a_N) - \lambda f_2(a_1, a_2, \ldots, a_N) \tag{48}$$

where

$$f_1 = I[u_A] \quad \text{and} \quad f_2 = J[u_A]$$

Requiring that U_T have a minimum leads to

$$\frac{\partial U_T}{\partial a_i} = \frac{\partial f_i}{\partial a_i} - \lambda \frac{\partial f_2}{\partial a_i} = 0 \quad \text{for} \quad i = 1, 2, \ldots, N \tag{49}$$

These equations are identical to Eqs. (26); thus, the Rayleigh-Ritz and the general Rayleigh-Timoshenko methods are identical. This is the reason that many authors call this particular application the Rayleigh-Ritz method as used by Timoshenko for buckling problems. Note that in this first type of application, the variation in $I - \lambda J$ with respect to u (keeping λ constant) leads to the same equations as the minimization of the Rayleigh quotient, given by Eq. (22).

Finally, for this type of application, convergence is guaranteed because we are dealing with a variational problem which satisfies the sufficiency conditions for a minimum. Some authors refer to this type of application as the Rayleigh-Ritz method, whereas when the method is applied to variational problems (stationary) that do not satisfy the sufficiency conditions for a minimum or a maximum, they call it simply the Ritz method. This distinction is not important. What is important is that there is no rigorous proof of convergence for this latter type of application, although the method has been used very successfully.

The second type of application does not depend on the existence of a Rayleigh quotient, and it is based on the stability criterion directly. If we express the deformation(s) by the finite series

$$u = \sum_{i=1}^{N} a_i g_i$$

where g_i are kinematically admissible functions, the total potential, $U_T[u]$ (functional) becomes a function of a_i, $U_T(a_i)$. For the equilibrium to be stable, the total potential must be a minimum, and the following conditions must be satisfied

$$\frac{\partial U_T}{\partial a_i} = 0 \quad i = 1, 2, \ldots, N \tag{50}$$

and

$$\begin{vmatrix} \dfrac{\partial^2 U_T}{\partial^2 a_1^2} & \dfrac{\partial^2 U_T}{\partial a_1 \partial a_2} & \cdots & \dfrac{\partial^2 U_T}{\partial a_1 \partial a_N} \\[2ex] \dfrac{\partial^2 U_T}{\partial a_2 \partial a_1} & \dfrac{\partial^2 U_T}{\partial a_2^2} & \cdots & \dfrac{\partial^2 U_T}{\partial a_2 \partial a_N} \\[2ex] \dfrac{\partial^2 U_T}{\partial a_N \partial a_1} & \dfrac{\partial^2 U_T}{\partial a_N \partial a_2} & \cdots & \dfrac{\partial^2 U_T}{\partial a_N^2} \end{vmatrix} > 0 \tag{51}$$

along with all its principal minors, such as

$$\frac{\partial^2 U_T}{\partial a_1^2} > 0, \qquad \begin{vmatrix} \dfrac{\partial^2 U_T}{\partial a_1^2} & \dfrac{\partial^2 U_T}{\partial a_1 \partial a_2} \\[2mm] \dfrac{\partial^2 U_T}{\partial a_2 \partial a_1} & \dfrac{\partial^2 U_T}{\partial a_2^2} \end{vmatrix} > 0, \qquad \text{etc.} \qquad (52)$$

Equations (50) give us the equilibrium equations that relate the load to the displacement parameters a_i (generalized coordinates). They are N equations in $N + 1$ unknowns (a_i, $i = 1, 2, \ldots, N$, and the load parameter λ). From these equations, we may solve for the a_i's in terms of the load parameter. Knowing the equilibrium positions, we then proceed to study the stability or instability of these equilibrium postions by using the inequalities given by Eqs. (51) and (52). The value of the load parameter at which the equilibrium changes from stable to unstable is the critical value. Note at this point that, if the expressions in Eqs. (51) and (52) are identically equal to zero, no decision can be made about the stability or instability of this equilibrium position, and higher variations are needed.

This procedure will be demonstrated in the following application. Consider a simply supported column of uniform geometry as shown in Fig. 5-3. The kinematic and constitutive relations are

$$\epsilon_{xx} = u_{,x} + \tfrac{1}{2}w_{,x}^2 - zw_{,xx}$$

$$\tau_{xx} = E\epsilon_{xx}$$

On the basis of these, the total potential is

$$U_T = \tfrac{1}{2} \int_0^L [EA(u_{,x} + \tfrac{1}{2}w_{,x}^2)^2 + EIw_{,xx}^2]\, dx + \bar{P}u(L) \qquad (53)$$

Let us now use the following one-term approximations for $u(x)$ and $w(x)$:

$$u(x) = B_1 x$$

$$w(x) = C_1 \sin\frac{\pi x}{L} \qquad (54)$$

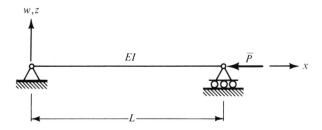

Figure 5-3. Geometry and sign convention for a simply supported column.

Note that $u(0) = 0$ and $w(0) = w(L) = 0$, and the functions x and $\sin \pi x/L$ are kinematically admissible. Substitution of Eqs. (54) into Eqs. (53) yields

$$U_T = \frac{L}{2}\left[EA\left\{B_1^2 + \frac{1}{2}\left(\frac{\pi}{L}\right)^2 B_1 C_1^2 + \frac{3}{32}\left(\frac{\pi}{L}\right)^4 C_1^4\right\} + \frac{1}{2}EI\left(\frac{\pi}{L}\right)^2 C_1^2\right] + \bar{P}LB_1$$

(55)

For equilibrium

$$\frac{\partial U_T}{\partial B_1} = \frac{\partial U_T}{\partial C_1} = 0$$

$$L\left[EAB_1 + \frac{EA}{4}\left(\frac{\pi}{L}\right)^2 C_1^2 + \bar{P}L\right] = 0$$

$$\frac{L}{2}\left(\frac{\pi}{L}\right)^2\left[EAB_1 C_1 + \frac{3}{8}EA\left(\frac{\pi}{L}\right)^2 C_1^3 + EI\left(\frac{\pi}{L}\right)^2 C_1\right] = 0$$

(56)

The second derivatives are given by

$$\frac{\partial^2 U_T}{\partial B_1^2} = EAL$$

$$\frac{\partial^2 U_T}{\partial B_1 \partial C_1} = \frac{EAL}{2}\left(\frac{\pi}{L}\right)^2 C_1$$

(57)

$$\frac{\partial^2 U_T}{\partial C_1^2} = \frac{EAL}{2}\left(\frac{\pi}{L}\right)^2 B_1 + \frac{9EAL}{16}\left(\frac{\pi}{L}\right)^4 C_1^2 + \frac{EIL}{2}\left(\frac{\pi}{L}\right)^4$$

If we let $P_E = \pi^2 EI/L^2$, the equilibrium equations, Eqs. (56), are

$$(EA)B_1 + \frac{EA}{4}\left(\frac{\pi}{L}\right)^2 C_1^2 = -\bar{P}$$

$$C_1\left[(EA)B_1 + \frac{3}{8}EA\left(\frac{\pi}{L}\right)^2 C_1^2 + P_E\right] = 0$$

(58)

It is easily seen from Eqs. (58) that there are two possible solutions:

(a) $\quad B_1 = -\dfrac{\bar{P}}{AE} \quad$ and $\quad C_1 = 0$

and

(b) $\quad B_1 = \dfrac{-\bar{P} - 2(\bar{P} - P_E)}{AE} \quad$ and $\quad C_1^2 = \dfrac{8}{AE}\left(\dfrac{L}{\pi}\right)^2 (\bar{P} - P_E)$

The corresponding deformation functions are

(a) $\quad u(x) = -\dfrac{\bar{P}}{AE}x, \qquad w(x) = 0$

(b) $\quad u(x) = -\dfrac{\bar{P} + 2(\bar{P} - P_E)}{AE}x$

and

$$w(x) = \pm \left(\frac{8}{AE}\right)^{1/2} \left(\frac{L}{\pi}\right) (\bar{P} - P_E)^{1/2} \sin \frac{\pi x}{L}$$

The term $(8/AE)^{1/2} (L/\pi) (\bar{P} - P_E)^{1/2}$ represents the maximum deflection, δ, (at $x = L/2$). From this, we may write the following two expressions for δ.

$$\frac{\delta}{L} = 2\sqrt{2} \left(\frac{\rho}{L}\right) \left(\frac{P}{P_E} - 1\right)^{1/2}$$

or

$$\frac{\delta}{\rho} = 2\sqrt{2} \left(\frac{P}{P_E} - 1\right)^{1/2}$$

where ρ is the radius of gyration of the cross-sectional area, $\rho^2 = I/A$. All of the equilibrium positions are shown, qualitatively, in Fig. 5-4.

The next problem is to determine the stability or instability of all the equilibrium positions. To this end, the two solutions are treated separately. First, let us consider the solution corresponding to the straight configuration $C_1 = 0$.

1. Making use of the expressions for the second partial derivatives, Eqs. (57), evaluated at $B_1 = -P/AE$ and $C_1 = 0$, we obtain the conditions for stability:

$$EAL > 0 \quad \text{and} \quad L\left(\frac{\pi}{L}\right)^2 [-\bar{P} + P_E] > 0 \tag{59}$$

It is clear from these inequalities that the straight configuration is stable for $\bar{P} < P_E$ and unstable for $\bar{P} > P_E$, as expected.

2. Similarly, since $EAL > 0$, the condition for stability for the equilibrium positions characterized by $C \neq 0$ is

$$\begin{vmatrix} EAL & \frac{EAL}{2}\left(\frac{\pi}{L}\right)^2 C_1 \\ \frac{EAL}{2}\left(\frac{\pi}{L}\right)^2 C_1 & \frac{\pi^2}{2L}\left[EAB_1 + \frac{9}{8}EA\left(\frac{\pi}{L}\right)^2 C_1^2 + P_E \right] \end{vmatrix} > 0$$

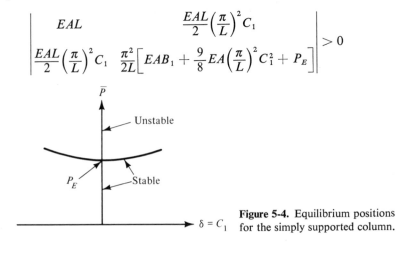

Figure 5-4. Equilibrium positions for the simply supported column.

Making use of the expressions for B_1 and C_1, and expanding the determinant, the above inequality becomes

$$\frac{\pi^2}{L}[\bar{P} - P_E] > 0$$

This inequality is definitely true, since these equilibrium positions (bent configuration) exist only if $P > P_E$.

The only question that remains to be answered is whether the position corresponding to $C_1 = 0$ and $\bar{P} = P_E$ (bifurcation point), is stable or unstable. If we expand the total potential about this position for small variations in B_1 and C_1, we obtain

$$\Delta U_T = \frac{1}{4!}\left(\frac{9}{8}\right)(EAL)\left(\frac{\pi}{L}\right)^4(\delta C_1)^4$$

Clearly, $\Delta U_T > 0$ and this position is stable.

In this particular application of the Rayleigh-Ritz method, it must be pointed out that the analysis is exact for the discrete system and it is approximate for the continuous system. The results obtained for the column are exact as far as the critical load is concerned, mainly because the exact form for $u(x)$ was assumed for the straight configuration, and because the expression for $w(x)$ is that of the linear eigenvalue problem. The results are also exact as far as the stability analysis is concerned. The only approximation involved is in the post-buckling curve for two reasons. First, because of the kinematic relations used, we do not expect the results to be applicable to bent configurations for which $w_{,x}^2 \ll 1$ does not hold. Second, even if $w_{,x}^2 \ll 1$, we do not necessarily have a good approximation for the bent configuration, and we must take more terms for the deformation functions to determine the convergence to the true deformation.

5.4   THE COLUMN BY THE TREFFTZ CRITERION

Consider the simply supported column, shown in Fig. 5-3. The kinematic and constitutive relations to be used are given by

$$\epsilon_{xx} = u_{,x} + \tfrac{1}{2}w_{,x}^2 - zw_{,xx}$$
$$= \epsilon_{xx}^0 - zw_{,xx} \tag{60}$$
$$\tau_{xx} = E\epsilon_{xx} \tag{61}$$

If we let

$$P = \int_A \tau_{xx}\, dA \tag{62}$$

where A is the cross-sectional area, the total potential is given by

$$U_T = \tfrac{1}{2} \int_0^L (P\epsilon_{xx}^0 + EIw_{,xx}^2)\, dx + \bar{P}u(L) \tag{63}$$

In terms of the displacements, the total potential becomes

$$U_T[u, w] = \tfrac{1}{2} \int_0^L [AE(u_{,x} + \tfrac{1}{2}w_{,x}^2)^2 + EIw_{,xx}^2]\, dx + \bar{P}u(L) \tag{64}$$

Let $\bar{u}(x)$ and $\bar{w}(x)$ denote positions of stable equilibrium, and let $\beta(x)$ and $\gamma(x)$ be kinematically admissible functions for $u(x)$ and $w(x)$, respectively. Then

$$U_T[\bar{u} + \epsilon_1\beta, \bar{w} + \epsilon_2\gamma] = \tfrac{1}{2} \int_0^L [AE\{\bar{u}_{,x} + \epsilon_1\beta_{,x} + \tfrac{1}{2}(\bar{w}_{,x} + \epsilon_2\gamma_{,x})^2\}^2$$
$$+ EI(\bar{w}_{,xx} + \partial_2\gamma_{,xx})^2]\, dx + \bar{P}[u(L) + \epsilon_1\beta(L)] \tag{65}$$

where ϵ_1 and ϵ_2 are small constants. After performing the indicated operations in the integrals and collecting like powers of the ϵ's, we have

$$U_T[\bar{u} + \epsilon_1\beta, \bar{w} + \epsilon_2\gamma] = \tfrac{1}{2} \int_0^L [EA(\bar{u}_{,x} + \tfrac{1}{2}\bar{w}_{,x}^2)^2 + EI\bar{w}_{,xx}^2]\, dx$$
$$+ \bar{P}u(L) + \epsilon_1 \left[\int_0^L EA(\bar{u}_{,x} + \tfrac{1}{2}\bar{w}_{,x}^2)\beta_{,x}\, dx + \bar{P}\beta(L) \right]$$
$$+ \epsilon_2 \int_0^L [EA(\bar{u}_{,x} + \tfrac{1}{2}\bar{w}_{,x}^2)\bar{w}_{,x}\gamma_{,x} + EI\bar{w}_{,xx}\gamma_{,xx}]\, dx$$
$$+ \frac{\epsilon_1^2}{2} \int_0^L EA\beta_{,x}^2\, dx + \epsilon_1\epsilon_2 \int_0^L EA\bar{w}_{,x}\beta_{,x}\gamma_{,x}\, dx \tag{66}$$
$$+ \frac{\epsilon_2^2}{2} \int_0^L [EA\bar{w}_{,x}^2\gamma_{,x}^2 + EA(\bar{u}_{,x} + \tfrac{1}{2}\bar{w}_{,x}^2)\gamma_{,x}^2 + EI\gamma_{,xx}^2]\, dx$$
$$+ \frac{\epsilon_1\epsilon_2^2}{2} \int_0^L EA\beta_{,x}\gamma_{,x}^2\, dx + \frac{\epsilon_2^3}{2} \int_0^L EA\bar{w}_{,x}\gamma_{,x}^3\, dx$$
$$+ \frac{\epsilon_2^4}{8} \int_0^L EA\gamma_{,x}^4\, dx$$

Note that the terms on the right side of Eq. (66) that do not contain ϵ's denote $U_T[\bar{u}, \bar{w}]$. Furthermore, if we collect terms with like powers in ϵ, we may write Eq. (66) in the following form:

$$\Delta U_T = \delta U_T + \delta^2 U_T + \delta^3 U_T + \delta^4 U_T \tag{67}$$

Next, if we are interested in studying the stability of equilibrium positions corresponding to the straight configuration

$$\Delta U_T[\bar{u} + \epsilon_1\beta, \epsilon_2\gamma] = \delta U_T[\bar{u} + \epsilon_1\beta, \epsilon_2\gamma] + \delta^2 U_T[\bar{u} + \epsilon_1\beta, \epsilon_2\gamma] + \cdots \quad (68)$$

where

$$\delta U_T[\bar{u} + \epsilon_1\beta, \epsilon_2\gamma] = \epsilon_1\left[\int_0^L P\beta_{,x}\, dx + \bar{P}\beta(L)\right]$$

$$\delta^2 U_T[\bar{u} + \epsilon_1\beta, \epsilon_2\gamma] = \frac{\epsilon_1^2}{2}\int_0^L EA\beta_{,x}^2\, dx + \frac{\epsilon_2^2}{2}\int_0^L [P\gamma_{,x}^2 + EI\gamma_{,xx}^2]\, dx \quad (69)$$

Note that $P = EA\bar{u}_{,x}$ from Eqs. (60)–(62).

Equilibrium for the straight configuration is characterized by

$$\delta U_T[\bar{u} + \epsilon_1\beta, \epsilon_2\gamma] = 0 \quad (70)$$

This leads to $P_{,x} = 0$ or $P = $ constant. Use of the boundary condition at $x = L$ yields $P = -\bar{P}$. These equilibrium positions are stable if $\delta^2 U_T$ is positive definite for all $\beta(x)$ and $\gamma(x)$ functions.

According to the Trefftz criterion (see Section 5.1), when the critical load is reached, $\delta^2 U_T$ becomes positive semidefinite. From the second equation of Eqs. (69), we notice that the first term is positive for all $\beta(x)$ except zero. Therefore the second term must be positive for all $\gamma(x)$ for stability. Thus, $\delta^2 U_T$ becomes positive semidefinite when $\beta(x) = 0$ and

$$\delta\int_0^L (P\gamma_{,x}^2 + EI\gamma_{,xx}^2)\, dx = 0$$

when $P = -\bar{P}$, or

$$\delta\int_0^L (EI\gamma_{,xx}^2 - \bar{P}\gamma_{,x}^2)\, dx = 0 \quad (71)$$

This condition leads to the same eigen-boundary-value problem as the one in Section 3.3. Note that the variations in Eq. (71) are with respect to kinematically admissible functions.

Alternate Procedure.   If we follow the approach used in Chapter 3, we notice that

$$\Delta U_T = \delta_\epsilon U_T = \int_0^L [P(\delta u_{,x} + w_{,x}\delta w_{,x} + \tfrac{1}{2}\delta w_{,x}^2)$$

$$+ EI(w_{,xx}\delta w_{,xx} + \tfrac{1}{2}\delta w_{,xx}^2)]\, dx + \bar{P}\delta u(L) = \int_0^L P\delta u_{,x}\, dx \quad (72)$$

$$+ \bar{P}\delta u(L) + \frac{1}{2}\int_0^L [P\delta w_{,x}^2 + EI\delta w_{,xx}^2]\, dx = \delta U_T + \delta^2 U_T$$

According to this approach, δU_T is the same as the first of Eqs. (69). The difference in $\delta^2 U_T$ between the two approaches [see the second of Eqs. (69)] is the term

$$\int_0^L \frac{\epsilon_1^2}{2} EA\beta(x) \, dx \tag{73}$$

This term, in the alternate approach, represents $\frac{1}{2}\int_0^L \delta P \, \delta u \, dx$ which is zero since the external and internal loads are kept constant during the virtual displacements δu and δw.

Therefore, again we have

$$\delta(\delta^2 U_T) = \delta\left[\int_0^L (EI\delta w_{,xx}^2 - \bar{P}\delta w_{,x}^2) \, dx\right] = 0$$

which is the same as Eq. (71).

It is important to note at this point that this particular form of the second variation is very attractive to the application of the Rayleigh-Ritz method, as demonstrated in the first type of application in Section 5.3.

$$\delta^2 U_T = I - \lambda J$$

where

$$I = \int_0^L EI\delta w_{,xx}^2 \, dx \tag{74}$$

$$\lambda = \bar{P}$$

$$J = \int_0^L \delta w_{,x}^2 \, dx$$

5.5   THE GALERKIN METHOD

The Galerkin method belongs in the class of approximate techniques for solving partial and ordinary differential equations. It was introduced in 1915 by B. G. Galerkin (Ref. 9) in the study of rods and plates, and it has been extensively used ever since by many investigators not only of problems in solid mechanics but also in fluid mechanics, heat transfer, and other fields. Finlayson and Scriven (Ref. 10) give an extensive bibliography on the uses of the Galerkin method. In addition, they unify this method with other approximate techniques under the name of the Method of Weighted Residuals (MWR).

Before outlining and applying the method to a number of problems, we must state that the method is not necessarily restricted to problems for which the differential equations are Euler-Lagrange equations (derived from stationary principles), and thus, this method is more general than the Rayleigh-Ritz technique. When dealing with variational problems, the Galerkin and

Ritz methods are closely related and under certain conditions completely equivalent (see Ref. 11).

The basic idea of the method is as follows: Suppose we require to solve the differential equation

$$L(u) = 0 \qquad 0 \le x \le L \tag{75}$$

where L is a differential operator, operating on u, which is a function of a single independent variable x, subject to some boundary conditions. We seek an approximate solution, u_{appr}, in the form

$$U_{appr} = \sum_{i=1}^{N} a_i f_i(x) \tag{76}$$

where $f_i(x)$ are a certain sequence of functions, each of which satisfies all of the boundary conditions, but none of them, as a rule, satisfy the differential equation, and a_i are undetermined coefficients. We can consider the functions to be elements of a complete sequence. If the exact solution to the differential equation, Eq. (75), is denoted by $\bar{u}(x)$, then the operator, L, operating on the difference $(u_{appr} - \bar{u})$ represents some kind of error or residual, $e(x)$,

$$e(x) = L(u_{appr} - \bar{u}) = L(u_{appr}) - L(\bar{u}) = L(u_{appr}) \tag{77}$$

If we substitute the series, Eq. (76), for u_{appr}, we have

$$e(x) = L\left(\sum_{i=1}^{N} a_i f_i(x)\right) \tag{78}$$

Next we must choose the undetermined coefficients, a_i, such that the error is a minimum. To this end, we make the error orthogonal, in the interval $0 \le x \le L$, to some weighting functions. In the Galerkin method the weighting functions are the functions used in the series, $f_k(x), k = 1, 2, \ldots, N$. This process leads to N integrals, called the Galerkin integrals

$$\int_0^L \left[L\left(\sum_{i=1}^{N} a_i f_i(x)\right) f_k(x) \right] dx = 0 \qquad k = 1, 2, \ldots, N \tag{79}$$

After performing the indicated operations, we have a system of N equations in N unknowns, a_i. To solution of this system is substituted into Eq. (76) to give the approximate solution to the problem. We obtain successive approximations by increasing N, and this gives us some idea about the convergence to the exact solution.

A number of questions and comments have been raised concerning choice of functions, convergence, and other particulars of the method. First, the choice of functions is not restricted by any means, but experience shows that, if the functions are elements of a complete sequence, convergence is improved.

Furthermore, which complete sequence must be used depends on the particular problem. When there are certain symmetries to be satisfied, if the functions are so chosen beforehand, it eliminates a lot of unnecessary work. As far as the boundary conditions are concerned, the method, as originally developed and applied by Galerkin, requires that the chosen functions satisfy all of the boundary conditions. This requirement can be relaxed, as will be shown in Section 5.5.1. This can easily be done for variational problems (when the differential equation is an Euler-Lagrange equation), but it presents difficulties in all other problems.

In variational problems we know precisely which boundary residuals or errors must be added and which must be subtracted from the Galerkin integral in order to relax the method. In nonvariational problems, the adding or subtracting of the boundary errors is based on mathematical convenience or the physics of the problem, and extreme care is required.

Second, convergence of the method has been and still is the subject of study for many mathematicians. Whenever the Galerkin and the Rayleigh-Ritz methods are equivalent, the convergence requirements and proofs for the Rayleigh-Ritz method imply convergence for the Galerkin method.

When the method is used in eigen-boundary-value problems, the Galerkin integrals lead to a system of N homogeneous algebraic equations in a_i. The requirement for a nontrivial solution leads to the vanishing of the determinant of the coefficients of the a_i, which is the characteristic equation.

5.5-1 The Method Derived From Stationary Principles

Although the Galerkin method may be used on all initial and boundary value problems, in the special case where it is applied to variational problems, it can be derived directly from the principle of the stationary value of the total potential. This is the case for all problems of elastostatics.

To demonstrate this, consider the beam-column problem, Fig. 3-2, treated in Chapter 3. Let us start with Eq. (15) of Chapter 3. For convenience, let us eliminate the in-plane component of deformation, through the use of the in-plane equilibrium equation ($P_{,x} = 0$, which implies that $P = $ constant, and from the boundary conditions $P = \bar{P}$). With this, Eq. (15) becomes

$$\int_0^L \left[(EIw_{,xx})_{,xx} - \bar{P}w_{,xx} - q(x) - \sum_{i=1}^n P_i\delta(x - x_i) + \sum_{j=1}^m C_j\eta(x - x_j) \right]\delta w\, dx$$

$$+ \left\{ \left[-(EIw_{,xx})_{,x} + \bar{P}w_{,x} \right]_{x=L} - R_L \right\}\delta w(L) - \left\{ \left[-(EIw_{,xx})_{,x} \right.\right.$$

$$\left.\left. + \bar{P}w_{,x} \right]_{x=0} - R_0 \right\}\delta w(0) + [(EIw_{,xx})_{x=L} - \bar{M}_L]\delta w_{,x}(L)$$

$$- [(EIw_{,xx})_{x=0}\bar{M}_0]\delta w_{,x}(0) = 0 \tag{80}$$

where δw denotes a virtual displacement.

From Eq. (80) we obtain the Euler-Lagrange equation and the associated boundary conditions.

D.E.

$$(EIw_{,xx})_{,xx} - \bar{P}w_{,xx} - q(x) - \sum_{i=1}^{n} P_i \delta(x - x_i) + \sum_{j=1}^{m} C_j \eta(x - x_j) = 0 \qquad (81)$$

Boundary Conditions

1. At $x = 0$

	Either	*or*
	$-(EIw_{,xx})_{,x} \bar{P}w_{,x} = R_0$	$\delta w = 0$
	$EIw_{,xx} = \bar{M}_0$	$\delta w_{,x} = 0$

2. At $x = L$

	Either	*or*
	$-(EIw_{,xx})_{,x} + \bar{P}w_{,x} = \bar{R}_L$	$\delta w = 0$
	$EIw_{,xx} = \bar{M}_L$	$\partial w_{,x} = 0$

Now let us suppose that for a given set of loads, $q(x)$, P_i, C_j, $\bar{P}$, we want to find the solution to the problem by employing Galerkin's method. We represent $w(x)$ by the series

$$w(x) = \sum_{m=1}^{N} a_m f_m(x) \qquad (83)$$

where $f_m(x)$ satisfy all of the boundary conditions regardless of whether they are kinematic or natural. Then, the Galerkin integrals are

$$\int_0^L \left[\left(EI \sum_{m=1}^{N} a_m f_{m,xx} \right)_{,xx} - \bar{P} \sum_{m=1}^{N} a_m f_{m,xx} - q(x) - \sum_{i=1}^{n} P_i \delta(x - x_i) \right.$$
$$\left. + \sum_{j=1}^{m} C_j \eta(x - x_j) \right] f_k \, dx = 0 \qquad k = 1, 2, 3, \ldots, N \qquad (84)$$

These are N linear algebraic equations in a_m ($m = 1, 2, \ldots, N$). We solve this system of equations for a_m, and we have the approximate solution by substituting these expressions for a_m into Eq. (83).

Another way of looking at the procedure is to directly associate it with Eq. (80). If the series representation for $w(x)$, Eq. (83), is substituted into Eq. (80), and if δw is taken to be $\delta a_k f_k(x)$, then we arrive at the same integrals as those given by Eqs. (84). Note that all the boundary terms vanish, and $\delta a_k \neq 0$ is taken outside the integral.

Next, suppose that the functions $f_m(x)$ in the series expressions for $w(x)$ satisfy only the kinematic boundary conditions. If we substitute the series into Eq. (80) we obtain

$$\delta a_k \int_0^L \left[\left(EI \sum_{m=1}^N a_m f_{m,xx} \right)_{,xx} - \bar{P} \sum_{m=1}^N a_m f_{m,xx} - q(x) - \sum_{i=1}^n P_i \delta(x - x_i) \right. $$

$$\left. + \sum_{j=1}^m C_j \eta(x - x_j) \right] f_k \, dx + \left\{ \left[-\left(EI \sum_{m=1}^N a_m f_{m,xx} \right)_{,x} \right. \right.$$

$$\left. + \bar{P} \sum_{m=1}^N a_m f_{m,x} \right]_{x=L} - \bar{R}_L \right\} \delta a_k f_k(L) - \left\{ \left[- EI \sum_{m=1}^N a_m f_{m,xx} \right)_{,x} \right.$$

$$\left. + \bar{P} \sum_{m=1}^N a_m f_{m,x} \right]_{x=0} - \bar{R}_0 \right\} \delta a_k f_k(0) + \left[\left(EI \sum_{m=1}^N a_m f_{m,xx} \right)_{x=L} \right.$$

$$\left. - \bar{M}_L \right] \delta a_k f_{k,x}(L) - \left[\left(EI \sum_{m=1}^N a_m f_{m,xx} \right)_{x=0} - \bar{M}_0 \right] \delta a_k f_{k,x}(0) = 0$$

$$k = 1, 2, \ldots, N \qquad (85)$$

As before, since $\delta a_k \neq 0$, Eqs. (85) represent a system of N linear algebraic equations in $a_m, m = 1, 2, \ldots, N$. The solution yields a_m and, therefore, the approximate expression for $w(x)$.

Note that in this modification of the Galerkin method we have added the boundary errors or residuals to the original Galerkin integrals. In addition, since we have related the method to the principle of the stationary value of the total potential, the functions $f_m(x)$ must be kinematically admissible. Among other requirements (see Appendix A), they must satisfy the kinematic boundary conditions.

5.5-2   The Clamped-Free Column

Consider a column of length L, which is clamped at $x = 0$ and free at $x = L$. This column is of uniform flexural stiffness, EI, and is loaded with axial load $\bar{P}$. We will use the Galerkin method to find P_{cr}. Let

$$w = \sum_{n=1}^3 a_n x^{n+1}$$

Since the functions x^{n+1} ($n = 1, 2, 3$) satisfy only the kinematic boundary conditions (at $x = 0$) and not the natural boundary conditions (at $x = L$), we will use the modified Galerkin method. Substitution of the above expression for $w(x)$ into Eq. (85) yields the following system of three homogeneous alebraic equations in a_n, $n = 1, 2, 3$. Wherever $\bar{P}$ appears, we must use $-\bar{P}$ because of the sign convention.

$$\int_0^L [EI24a_3 + \bar{P}(2a_1 + 6a_2x + 12a_3x^2)] \begin{bmatrix} x^2 \\ x^3 \\ x^4 \end{bmatrix} dx + [-EI(6a_2 + 24a_3L)$$

$$\bar{P}(2a_1L + 3a_2L^2 + 4a_3L^3)] \begin{bmatrix} L^2 \\ L^3 \\ L^4 \end{bmatrix} + EI(2a_1 + 6a_2L + 12a_3L^2) \begin{bmatrix} 2L \\ 3L^2 \\ 4L^3 \end{bmatrix}$$

$$= 0 \tag{86}$$

If we perform the indicated operations, divide through by EI, and introduce the load parameter $\lambda = \bar{P}L^2/EI$, we obtain the following three equations

$$4(1 - \tfrac{1}{3}\lambda)a_1 + 3L(2 - \tfrac{1}{2}\lambda)a_2 + 8L^2(1 - \tfrac{1}{5}\lambda)a_3 = 0$$
$$3(2 - \tfrac{1}{2}\lambda)a_1 + 3L(4 - \tfrac{3}{5}\lambda)a_2 + 2L^2(9 - \lambda)a_3 = 0 \tag{87}$$
$$8(1 - \tfrac{1}{5}\lambda)a_1 + 2L(9 - \lambda)a_2 + 16L^2(\tfrac{9}{5} - \tfrac{1}{7}\lambda)a_3 = 0$$

These equations are similar to Eqs. (30) obtained by the Rayleigh-Timoshenko or Rayleigh-Ritz method, and the solution is identical to the one obtained, or

$$\begin{array}{lll} \text{one-term solution} & \lambda_{cr} = 3 \\ \text{two-term solution} & \lambda_{cr} = 2.4860 \\ \text{three-term solution} & \lambda_{cr} = 2.4677 \end{array}$$

5.6 SOME COMMENTS ON KOITER'S THEORY

5.6-1 Critical Load and Load-Carrying Capacity

As mentioned in Chapter 1, the interest in the stability of simple structural elements or overall structural configurations under external causes lies in the fact that the stability limit (critical load) in many cases forms the basic criterion for design. Because of safety reasons in designing structural configurations, the level of the external causes is usually kept at such a value that the load in the structural configurations is smaller than the critical load or condition. This line of thinking might suggest that there is no reason to concern ourselves with studies of how the structural configuration behaves past this critical condition, because the critical condition is directly associated with the load-carrying capacity of the structural configuration. It has been known since the last century, though, that certain structural configurations (the rectangular, simply supported plate under uniform edge compression, the simply supported or clamped plate under uniform radial compression around its circumference, and others) can carry loads higher than the first buckling load (and still behave elastically in many cases). This fact has been verified experimentally. In a number of other cases it has been demonstrated experimentally (thin spherical shells under uniform compression and thin circular cylindrical shells under uniform compression) that the buckling load is only a small fraction of the critical load predicted by the mathematical model, based on either the equilibrium approach or the energy criterion. Many attempts have been made to explain this discrepancy, and it is beyond the scope of this text to go into such explanations. What is important, though, is the observation that critical loads, derived on the basis that the primary path becomes unstable when a bifurcation point exists, cannot be directly associated with the load-carrying capacity of such structural configurations. Note that, at a point of bifurcation, the branch that characterizes the adjacent equilibrium positions can be stable or unstable (see Figs. 1-4 and 1-5). Therefore, in the interest of designing a safe structure, we must know how the structural configuration behaves past the critical load or condition. The first person to systematically develop a stability theory which deals with the question of post-buckling behavior of continuous elastic systems is Koiter in his famous Ph.D. Thesis in 1945 (Ref.2).

His theory is an initial post-buckling analysis and therefore it cannot possibly provide all the required answers in relating the critical load to the load-carrying capacity of the structural configuration. Although it has this limitation, it is an important first step toward achieving the true solution. These points will be discussed in more detail with general qualitative demonstrations in Section 5.6.2.

The initial post-buckling behavior of elastic systems has received the deserved attention only in the past fifteen years. An excellent review article on the subject was presented by Hutchinson and Koiter (Ref. 12) in 1970.

5.6-2 Conclusions Based on Koiter's Theory

Koiter's theory, as mentioned previously, is primarily an initial post-buckling theory, and it is applicable to problems exhibiting bifurcational buckling only. In addition, the theory as presented in Ref. 2 is limited to linearly elastic behavior. It first concerns itself with the investigation of the equilibrium position in the neighborhood of the buckling load. The most important conclusion of this investigation is that the stability or instability of these equilibrium positions is governed by the stability of equilibrium at the bifurcation point. If the equilibrium is stable at the critical load (bifurcation point), neighboring positions of equilibrium can exist only for loads greater than the critical load, and these positions are stable (see Fig. 2-12a). If the equilibrium position at the critical load is unstable, neighboring equilibrium positions do exist at loads smaller than the critical load and they are unstable (see Figs. 2-12b and 2-12c). It is true that, in this particular case, it is possible for stable equilibrium positions also to exist at loads greater than the critical load (see Fig. 2-12c), but these positions can be reached only by passing through the unstable critical position, and therefore their practical significance is, to say the least, doubtful.

Another important ingredient of Koiter's theory is the investigation of the influence of small imperfections in the actual structure in comparison to the idealized perfect model. The most important conclusion of this part of Koiter's work is that, if the equilibrium position at the critical load of the perfect model is unstable (Figs. 2-12b and 2-12c), the critical load of the structural configuration may be considerably smaller than that of the idealized perfect model because of the presence of small imperfections (see Fig. 2-11).

Stein (Ref. 13), in a review paper, suggests that the Koiter theory can serve to assess the imperfection sensitivity of a structural configuration through the slope of the post-buckling curve in the neighborhood of the critical load. In a plot of normalized load versus normalized characteristic displacement, Fig. 5-5, the pre-buckling curve for linear theory is a 45° straight line. If we introduce an angle φ between the horizontal and the tangent to the post-buckling curve (positive in a counterclockwise direction), then it is clear that if $0 < \varphi < \pi/4$, the structure is imperfection insensitive, while if $-3\pi/4 < \varphi < 0$, the structure is imperfection sensitive. Two questions arise for this latter case. First, how large must the negative angle φ be to distinguish the cases of small versus large effects of imperfection sensitivity? Second, by considering the tangent to the post-buckling curve at the critical load, do we have assurance that the effect of imperfection sensitivity is considerably large? Suppose that there is a possibility that two structural

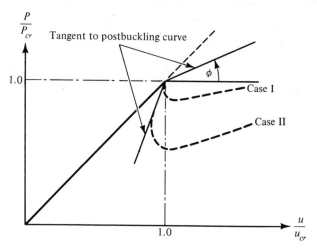

Figure 5-5. Initial postbuckling behavior.

configurations are characterized by the same tangent at the critical load but their behavior differs considerably as we move away from the critical load (see Fig. 5-5, cases I and II). These questions and others with the assessment of the effect of imperfection sensitivity are the subject of present-day research. Again the reader is strongly advised to read Ref. 12.

PROBLEMS

In all of these problems, use one of the approximate methods discussed in this chapter.

1. Find P_{cr} for a column which is fixed at one end and simply supported at the other.

2. A continuous column of constant flexural stiffness and total length $3L$ is supported as shown. Find P_{cr}.

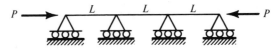

Figure P5-2.

3. A column of constant flexural stiffness, EI, and length, L, is fixed at one end and supported elastically against translation (linear spring) at the other. If the column is loaded axially by P, find P_{cr}.

4. A cantilever column of constant flexural stiffness, EI, and length, L, is in an upright position with the fixed end at the lower part. If the direction of gravity is in line with the column, determine the critical weight if the column is to buckle under its own weight.

5. A simply supported column of length $3L$ is under the action of a compressive load. Find P_{cr} if the flexural stiffness varies according to

$$EI(x) = \begin{cases} EI_0 & 0 \le x \le L \\ 2EI_0 & L \le x \le 2L \\ EI_0 & 2L \le x \le 3L \end{cases}$$

6. A cantilever column of length $2L$ is fixed at $x = 0$ and loaded at the free end by a compressive load P. Find P_{cr} if the flexural stiffness varies according to

$$EI(x) = \begin{cases} 2EI_0 & 0 \le x \le L \\ EI_0 & L \le x \le 2L \end{cases}$$

7. A cantilever column, with the clamped end at $x = L$ is under a compressive load P (at the free end, $x = 0$). Find P_{cr} if the the stiffness varies according to $EI(x) = EI_0(1 + I_1/I_0 x)$.

REFERENCES

1. SAGAN, H., *Introduction to the Calculus of Variations*, McGraw-Hill Book Co., New York, 1969.

2. KOITER, W. T., "Elastic Stability and Postbuckling Behavior," in *Non-linear Problems*, edited by R. E. Langer, University of Wisconsin Press, Madison, 1963. Also "The Stability of Elastic Equilibrium," Thesis Delft, 1945 (English translation, NASA TT-F-10833, 1967, and AFFDL TR-70-25, 1970).

3. TIMOSHENKO, S. P., and GERE, J. M., *Theory of Elastic Stability*, McGraw-Hill Book Co., New York, 1956.

4. TREFFTZ, E., "Zur Theorie der Stabilität des Elastichen Gleichogewihts," *Z. Angew. Math. Mech.*, Vol. 13, pp. 160–165, 1933.

5. RAYLEIGH, J. W. S., *Theory of Sound*, Dover Publications, New York, 1945.

6. TADJBAKHSH, I., and KELLER, J. B., "Strongest Columns and Isoperimetric Inequalities for Eigen Values," *J. Appl. Mech.*, Vol. 29, pp. 159–164, 1962.

7. PRAGER, W., and TAYLOR, J. E., "Problem of Optimal Structural Design," *J. Appl. Mech.*, Vol. 35, pp. 102–106, 1968.

8. SIMITSES, G. J., KAMAT, M. P., and SMITH, C. V., JR., "The Strongest Column by the Finite Element Displacement Method," AIAA Paper No. 72-141, 1972.

9. GALERKIN, B. G., "Sterzhnei i plastiny. Ryady V Nekotorykh Voprosakh Uprogogo Ravnoresiya Sterzhnei i plastiny," (Rods and Plates. Series Occurring in Some Problems of Elastic Equilibrium of Rods and Plates), *Vestn. Inzhen, i Tekh. Petrograd*, Vol. 19, pp. 897–908, 1915; English translation 63–18924 Clearinghouse Fed. Sci. Tech. Info. See also "On the Seventieth Anniversary of the Birth of B.G. Galerkin," *PMM*, Vol. 5, pp. 337–341, 1941.

10. FINLAYSON, B. A., and SCRIVEN L. E., "The Method of Weighted Residuals—A Review," *Appl. Mech. Rev.*, Vol. 19, No. 9, pp. 735–748, 1966.

11. SINGER, J., "On the Equivalence of the Galerkin and Rayleigh-Ritz Methods," *Journal of Royal Aerospace Society*, Vol. 66, pp. 592–597, 1962.

12. HUTCHINSON, J. W., and KOITER, W. T., "Postbuckling Theory," *Appl. Mech. Rev.*, Vol. 13, pp. 1353–1366, 1970.

13. STEIN, M., "Recent Advances in Shell Buckling," AIAA Paper No. 68-103, 1968.

6

COLUMNS

ON ELASTIC FOUNDATIONS

6.1 BASIC CONSIDERATIONS

Beams and columns supported elastically along their lengths are widely found in structural configurations. In some cases, the elastic support, called the elastic foundation, is provided by a medium which is indeed the foundation supporting the beams or columns such as in railroad tracks, in underground piping for different uses, and in footings for large-scale structures. In other cases, the elastic support is provided by adjacent elastic structural elements such as in stiffened plate and shell configurations. Regardless of the particular application, the mathematical model consists of a column supported in some manner at its ends and with a continuous distribution of springs of stiffness, $\bar{\beta}$, called the modulus of the foundation (see Fig. 6-1). The units of $\bar{\beta}$ are pounds per inch per inch (force per length squared) and may be a constant or at most a function of position along the length of the column for linear spring behavior. In general, the spring behavior may be taken as nonlinear.

This chapter will present the analysis of some simple models and provide insight into the behavior of such columns under destabilizing compressive loads. An excellent and comprehensive treatment of the subject may be found in the text by Heteńyi (Ref. 1).

To derive the buckling equations for a column on an elastic foundation, we refer to Chapter 3 and modify the expression for the total potential to include the energy stored into the foundation, U_f. For linear spring behavior

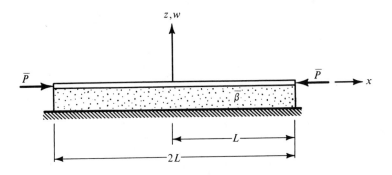

Figure 6-1. Column resting on an elastic foundation.

$$U_f = \tfrac{1}{2} \int_0^L \bar{\beta} w^2 \, dx \tag{1}$$

Therefore, if we use the principle of the stationary value of the total potential

$$\delta_\epsilon U_T = \delta_\epsilon U_i + \delta_\epsilon U_f + \delta_\epsilon U_p = 0 \tag{2}$$

we can derive the equilibrium equations for this configuration. Since $P = $ constant, the buckling equation is

$$(EI w \cdot_{xx})_{,xx} + \bar{P} w_{,xx} + \bar{\beta} w = 0 \tag{3}$$

The boundary conditions are not affected by the presence of the foundation.

6.2   THE PINNED-PINNED COLUMN

Consider a column of length, L, and constant flexural stiffness, EI, pinned at both ends and resting on an elastic foundation (see Fig. 6-2). Let the modulus of the foundation, $\bar{\beta}$, be a constant. The mathematical formulation of the problem is given by

D.E. $\quad w_{,xxxx} + k^2 w_{,xx} - \left(\dfrac{\pi}{L}\right)^4 \beta w = 0 \tag{4a}$

B.C.'s $\quad w(0) = 0 \qquad w(L) = 0$

$\qquad\qquad w_{,xx}(0) = 0 \qquad w_{,xx}(L) = 0 \tag{4b}$

where

$$k^2 = \frac{P}{EI} \quad \text{and} \quad \beta = \frac{\bar{\beta} L^4}{\pi^4 EI}$$

Thus, the problem has been reduced to an eigen-boundary-value problem and we are seeking the smallest value of $k(P_{cr})$ for which a nontrivial solution exists provided that β is fixed.

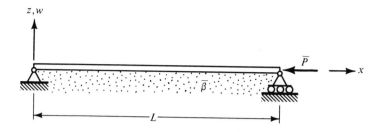

Figure 6-2. The pinned-pinned column.

Since $\sin m\pi x/L$ satisfies all boundary conditions for all m, the solution to the buckling equation is taken in the form

$$w(x) = \sum_{m=1}^{\infty} A_m \sin \frac{m\pi x}{L} \tag{5}$$

Substitution into the differential equations results in the following characteristic equation:

$$\left(\frac{m\pi}{L}\right)^4 - k_m^2 \left(\frac{m\pi}{L}\right)^2 + \beta \left(\frac{\pi}{L}\right)^4 = 0 \tag{6}$$

where k_m denote the eigenvalues.

From Eq. (6) we obtain

$$k^2 \left(\frac{L}{\pi}\right)^2 = \frac{P_m}{P_E} = m^2 + \frac{\beta}{m^2} \tag{7}$$

where

$$P_E = \frac{\pi^2 EI}{L^2}$$

The critical load for a fixed value of the modulus of the foundation, β, is the smallest of P_m. We see from Eq. (7) that P_{cr} and the corresponding deformation mode depend on the value of β. Thus, P_{cr} is obtained from a plot of P_m/P_E versus β (see Fig. 6-3). As shown in the plot, P_{cr} is denoted by the solid piecewise linear curve and

$$P_{cr} = P_1 \qquad 0 \leq \beta \leq 4$$
$$P_{cr} = P_2 \qquad 4 \leq \beta \leq 36$$
$$\text{etc.}$$

In general, the value of β at which the deformation mode changes from k half-sine waves to $(k + 1)$, and the critical load from P_k to P_{k+1}, is given by

$$\beta = [k(k + 1)]^2 \tag{8}$$

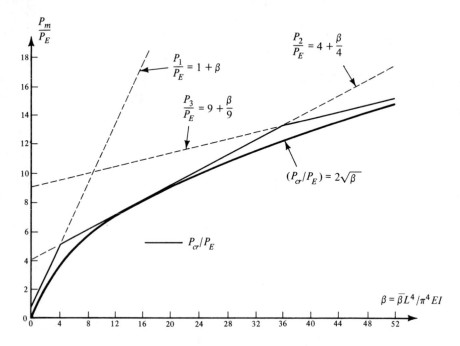

Figure 6-3. Critical conditions for the pinned-pinned column.

Finally, if we consider m^2 to be a continuous variable, minimization of P_m with respect to m^2 yields

$$m_{cr}^2 = \sqrt{\beta} \tag{9}$$

Substitution of this expression into Eq. (7) gives us the expression for the critical load:

$$P_{cr} = 2P_E\sqrt{\beta} = 2\sqrt{\bar{\beta}EI} \tag{10}$$

This expression is also plotted in Fig. 6-3. We notice that this form of the solution is approximate, and the approximation becomes more accurate as β increases.

6.3   RAYLEIGH-RITZ SOLUTION

Using Timoshenko's arguments about energy and work, we can derive a Rayleigh quotient for this problem:

$$P = \frac{\int_0^L EIw_{,xx}^2 \, dx + \int_0^L \bar{\beta}w^2 \, dx}{\int_0^L w_{,x}^2 \, dx} \tag{11}$$

The solution may be expressed in terms of the following series for the pinned-pinned column:

$$w(x) = \sum_{m=1}^{N} A_m \sin \frac{m\pi x}{L} \tag{12}$$

When this expression is used in Eq. (11), the numerator and denominator become functions of A_m:

$$P = \frac{f_1(A_m)}{f_2(A_m)} \tag{13}$$

Next, we adjust the coefficient A_m such that P_m is a minimum. This leads to the equation

$$\frac{\partial f_1}{\partial A_m} - P\frac{\partial f_2}{\partial A_m} = 0 \qquad m = 1, 2, \ldots, N \tag{14}$$

which becomes

$$L\left[EI\left(\frac{m\pi}{L}\right)^4 + \bar{\beta} - \bar{P}\left(\frac{m\pi}{L}\right)^2\right]A_m = 0 \qquad m = 1, 2, \ldots, N \tag{15}$$

Thus, the introduction of P_E and β yields Eq. (7), and from this point on the arguments are the same as in Section 6.2.

6.4 THE GENERAL CASE

In this section we shall outline a procedure which may, in general, be applied to columns on an elastic foundation regardless of the boundary conditions. Consider a column of constant flexural stiffness and length $2L$ as shown in Fig. 6-1. Place the origin of the coordinate system used at the midpoint of the column.

The mathematical formulation of the general problem is given by

D.E.
$$w_{,xxxx} + \frac{\bar{P}}{EI}w_{,xx} + \frac{\bar{\beta}}{EI}w = 0 \tag{16}$$

B.C.'s   *Either* (Kinematic)   *Or* (Natural)

$$w = 0 \quad \text{at} \quad x = \pm L \qquad w_{,xxx} + \frac{\bar{P}}{EI}w_{,x} = 0 \quad \text{at} \quad x = \pm L$$

$$w_{,x} = 0 \quad \text{at} \quad x = \pm L \qquad w_{,xx} = 0 \quad \text{at} \quad x = \pm L$$

Note that the clamped-clamped problem is characterized by the satisfaction of the kinematic boundary conditions, whereas the free-free problem (floating ends) is characterized by the satisfaction of the natural boundary conditions, etc.

The general procedure is as follows. Let the solution to Eq. (16) be of the form e^{sx}. Then, substitution into Eq. (16) yields

$$s^4 + \frac{P}{EI}s^2 + \frac{\bar{\beta}}{EI} = 0 \qquad (17)$$

The solution for s is given by

$$s^2 = \frac{1}{2}\left[-\frac{\bar{P}}{EI} \pm \sqrt{\left(\frac{\bar{P}}{EI}\right)^2 - 4\frac{\bar{\beta}}{EI}}\right]$$

$$= \left(\frac{\bar{\beta}}{EI}\right)^{1/2}\left[-\frac{\bar{P}}{2\sqrt{\bar{\beta}EI}} \pm \sqrt{\left(\frac{\bar{P}}{2\sqrt{\bar{\beta}EI}}\right)^2 - 1}\right]$$

Denoting the expression $2\sqrt{\bar{\beta}EI}$ by P_1, which represents P_{cr} for large values of β (not $\bar{\beta}$) or extremely long columns with simply supported boundaries, then

$$s^2 = \left(\frac{\bar{\beta}}{EI}\right)^{1/2}\left[-\frac{\bar{P}}{P_1} \pm \sqrt{\left(\frac{\bar{P}}{P_1}\right)^2 - 1}\right] \qquad (18)$$

Furthermore, if we let $\bar{P}/P_1 = \gamma$, the four roots of Eq. (17) are

$$s_1 = i\left(\frac{\bar{\beta}}{EI}\right)^{1/4}[\gamma - \sqrt{\gamma^2 - 1}]^{1/2}$$

$$s_2 = -s_1$$

$$s_3 = i\left(\frac{\bar{\beta}}{EI}\right)^{1/4}[\gamma + \sqrt{\gamma^2 - 1}]^{1/2} \qquad (19)$$

$$s_4 = -s_3$$

Regardless of the boundary conditions, the following three cases must be considered.

Case I:

$$\gamma > 1$$

For this case, let us first introduce the real and positive parameters k_1 and k_2:

$$k_1 = \left(\frac{\bar{\beta}}{EI}\right)^{1/4}[\gamma - \sqrt{\gamma^2 - 1}]^{1/2}$$

$$k_2 = \left(\frac{\bar{\beta}}{EI}\right)^{1/4}[\gamma + \sqrt{\gamma^2 - 1}]^{1/2} \qquad (20)$$

Then the four roots and the solution to Eq. (16) become

$$s_1 = ik_1 \qquad s_2 = -ik_1 \qquad s_3 = ik_2 \qquad s_4 = -ik_2$$

and

$$w(x) = C_{11} \cos k_1 x + C_{12} \sin k_1 x + C_{13} \cos k_2 x + C_{14} \sin k_2 x \qquad (21)$$

Case II:

$$\gamma = 1$$

For this case, we first introduce the real positive parameter k_3:

$$k_3 = \left(\frac{\bar{\beta}}{EI}\right)^{1/4} \qquad (22)$$

Then, the four roots and the solution to Eq. (16) become

$$s_1 = ik_3 \qquad s_2 = -ik_3 \qquad s_3 = ik_3 \qquad s_4 = -ik_3$$

and

$$w(x) = C_{21} \cos k_3 x + C_{22} \sin k_3 x + C_{23} x \cos k_3 x + C_{24} x \sin k_3 x \qquad (23)$$

Note that we have two pairs of double roots for this case.

Case III:

$$\gamma < 1$$

Since γ is smaller than 1, then the four roots, Eqs. (19), are

$$s_1 = i\left(\frac{\bar{\beta}}{EI}\right)^{1/4} [\gamma - i\sqrt{1 - \gamma^2}]^{1/2}$$

$$s_2 = -s_1$$

$$s_3 = i\left(\frac{\bar{\beta}}{EI}\right)^{1/4} [\gamma + i\sqrt{1 - \gamma^2}]^{1/2}$$

$$s_4 = -s_3$$

If we take the square root of the complex number and introduce the real positive quantities

$$p = \left(\frac{\bar{\beta}}{EI}\right)^{1/4} \sqrt{\frac{1 - \gamma}{2}}$$

$$r = \left(\frac{\bar{\beta}}{EI}\right)^{1/4} \sqrt{\frac{1 + \gamma}{2}} \qquad (24)$$

the four roots become

$$s_1 = p + ir = \eta$$
$$s_2 = -p - ir$$
$$s_3 = -p + ir = \omega \tag{24b}$$
$$s_4 = p - ir$$

Note that we have two complex conjugate pairs $s_4 = \bar{s}_1$ and $s_2 = \bar{s}_3$. The solution to Eq. (16) for this case is

$$w(x) = A_1 e^{(p+ir)x} + A_2 e^{-(p+ir)x} + A_3 e^{(-p+ir)x} + A_4 e^{(p-ir)x}$$

or

$$w(x) = C_{31} \cosh \eta x + C_{32} \cosh \omega x + C_{33} \sinh \eta x + C_{34} \sinh \omega x \tag{25}$$

6.4-1 The Clamped-Clamped Column

To find P_{cr} for this particular problem, we must investigate all three cases by using the proper boundary conditions with the three corresponding solutions, Eqs. (21), (23), and (25). Note that in all three cases, when the boundary conditions are used, we end up with a system of four linear homogeneous algebraic equations in the constants C_{ij} ($i = 1, 2, 3; j = 1, 2, 3, 4$). The first subscript, i, is associated with the particular case (I, II, and III) and the second subscript, j, with the four roots.

The boundary conditions for the clamped-clamped case are

$$w(-L) = w(L) = 0$$
$$w_{,x}(-L) = w_{,x}(L) = 0 \tag{26}$$

Case III:

$$\gamma < 1$$

Employing the boundary conditions, Eqs. (26), and the expression for $w(x)$, Eq. (25), we obtain the following system of equations:

$$C_{31} \cosh \eta L + C_{32} \cosh \omega L \pm C_{33} \sinh \eta L \pm C_{34} \sinh \omega L = 0$$
$$\pm C_{31}\eta \sinh \eta L \pm C_{32}\omega \sinh \omega L + C_{33}\eta \cosh \eta L + C_{34}\omega \cosh \omega L = 0 \tag{27}$$

If we add and subtract the first two equations and the last two equations, we have an equivalent system of four linear homogeneous algebraic equations in C_{3j} ($j = 1, 2, 3,$ and 4):

$$C_{31} \cosh \eta L + C_{32} \cosh \omega L = 0$$
$$C_{33} \sinh \eta L + C_{34} \sinh \omega L = 0$$
$$C_{31}\eta \sinh \eta L + C_{32}\omega \sinh \omega L = 0$$
$$C_{33}\eta \cosh \eta L + C_{34}\omega \cosh \omega L = 0$$

Thus, the equations have decomposed into two systems of equations:

$$\text{Either} \qquad \left. \begin{array}{l} C_{31} \cosh \eta L + C_{32} \cosh \omega L = 0 \\ C_{31}\eta \sinh \eta L + C_{32}\omega \sinh \omega L = 0 \end{array} \right\} \tag{28}$$

$$\text{Or} \qquad \left. \begin{array}{l} C_{33} \sinh \eta L + C_{34} \sinh \omega L = 0 \\ C_{33}\eta \cosh \eta L + C_{34}\omega \cosh \omega L = 0 \end{array} \right\} \tag{29}$$

The first system implies that $C_{31} \neq 0$, $C_{32} \neq 0$, and $C_{33} = C_{34} = 0$, which corresponds to a symmetric mode of deformations [see Eq. (25)]. The second system corresponds to an antisymmetric mode of deformation. Both systems must be used for finding P_{cr}.

For the symmetric case, a nontrivial solution exists if

$$\begin{vmatrix} \cosh \eta L & \cosh \omega L \\ \eta \sinh \eta L & \omega \sinh \omega L \end{vmatrix} = 0 \tag{30}$$

The expansion of the determinant yields

$$\omega \cosh \eta L \sinh \omega L - \eta \sinh \eta L \cosh \omega L = 0 \tag{31}$$

This equation may now be written as

$$\omega(e^{\eta L} + e^{-\eta L})(e^{\omega L} - e^{-\omega L}) - \eta(e^{\eta L} - e^{-\eta L})(e^{\omega L} + e^{-\omega L}) = 0$$

If we substitute the expressions for η and ω from Eqs. (24b), we obtain

$$4i[p \sin 2rL + r \sinh 2pL] = 0 \tag{32a}$$

or

$$\frac{\sin 2rL}{2rL} + \frac{\sinh 2pL}{2pL} = 0 \tag{32b}$$

This equation has no solution, therefore $C_{31} = C_{32} = 0$ (trivial solution for the system).

Similarly, for the antisymmetric case, the characteristic equation requires

$$\frac{\sin 2rL}{2rL} - \frac{\sinh 2pL}{2pL} = 0 \tag{33}$$

This equation has no solution for γ; therefore, for this case ($\gamma < 1$), there is no bifurcation point and the only solution is $w(x) \equiv 0$ (straight configuration).

Case II:

$$\gamma = 1$$

If the steps outlined for case III are repeated for this case using Eq. (23) for the displacement, we obtain the following characteristic equations for symmetric buckling ($C_{21} \neq 0, C_{24} \neq 0, C_{22} = C_{23} = 0$):

$$\sin 2k_3 L = -2k_3 L \qquad (34a)$$

and for antisymmetric buckling ($C_{21} = C_{24} = 0, C_{22} \neq 0, C_{23} \neq 0$):

$$\sin 2k_3 L = 2k_3 L \qquad (34b)$$

There is no solution to Eqs. (34); therefore, there is no bifurcation for $\gamma = 1$.

Case I:
Substitution of Eq. (21) into the boundary conditions, Eqs. (26), yields

$$C_{11} \cos k_1 L \pm C_{12} \sin k_1 L + C_{13} \cos k_2 L \pm C_{14} \sin k_2 L = 0$$
$$\mp C_{11}k_1 \sin k_1 L + C_{12}k_1 \cos k_1 L \mp C_{13}k_2 \sin k_2 L + C_{14}k \cos k_2 L = 0 \qquad (35)$$

As in case III we first obtain an equivalent system of equations through subtraction and addition, which separates the problem into symmetric and antisymmetric buckling.

For symmetric buckling ($C_{11} \neq 0, C_{13} \neq 0, C_{12} = C_{14} = 0$):

$$C_{11} \cos k_1 L + C_{13} \cos k_2 L = 0$$
$$C_{11}k_1 \sin k_1 L + C_{13}k_2 \sin k_2 L = 0 \qquad (36)$$

This leads to the characteristic equation

$$k_2 \sin k_2 L \cos k_1 L = k_1 \sin k_1 L \cos k_2 L$$

or

$$(k_1 L) \tan (k_1 L) = (k_2 L) \tan (k_2 L) \qquad (37)$$

For antisymmetric buckling ($C_{11} = C_{13} = 0, C_{12} \neq 0, C_{14} \neq 0$):

$$C_{12} \sin k_1 L + C_{14} \sin k_2 L = 0$$
$$C_{12}k_1 \cos k_1 L + C_{14}k_2 \cos k_2 L = 0 \qquad (38)$$

The characteristic equation is

$$(k_1L) \cot (k_1L) = (k_2L) \cot (k_2L) \tag{39}$$

To find γ_{cr}, we must solve both Eq. (37) and Eq. (39) for fixed values of $\bar{\beta}$. In Ref. 1, Heteñyi presents graphically the solution to the two characteristic equations in a plot of γ versus $4L^2\sqrt{\bar{\beta}/EI}$. In the same reference, plots for the pinned-pinned and free-free columns are presented with the same coordinates. For the clamped-clamped and pinned-pinned columns, as $\bar{\beta}$ is increased from zero, the buckling mode changes from symmetric to antisymmetric back to symmetric, etc. For the free-free column, as $\bar{\beta}$ is increased from zero, the buckling mode changes from antisymmetric to symmetric, etc.

6.4-2 The Free-Free Column

The boundary conditions for this particular problem are:

$$w_{,xx}(-L) = w_{,xx}(L) = 0$$
$$w_{,xxx}(-L) + \frac{\bar{P}}{EI}w_{,x}(-L) = w_{,xxx}(L) + \frac{\bar{P}}{EI}w_{,x}(L) = 0 \tag{40}$$

To find $\bar{P}_{cr}$, we must again consider all three cases. It can be shown that no solution exists for cases I and II ($\gamma > 1$, $\gamma = 1$; see Problem 1 at the end of this chapter). Therefore, if there is a $\bar{P}_{cr}$, the characteristic equation must be found from case III ($\gamma < 1$). Substitution of Eq. (25) into the boundary conditions, Eqs. (40), results in:

$$C_{31}\eta^2 \cosh \eta L + C_{32}\omega^2 \cosh \omega L \mp C_{33}\eta^2 \sinh \eta L \mp C_{34}\omega^2 \sinh \omega L = 0$$
$$[\mp C_{31}\eta^3 \sinh \eta L \mp C_{32}\omega^3 \sinh \omega L + C_{33}\eta^3 \cosh \eta L + C_{34}\omega^3 \cosh \omega L]$$
$$+ \frac{\bar{P}}{EI}[\mp C_{31}\eta \sinh \eta L \mp C_{32}\omega \sinh \omega L + C_{33}\eta \cosh \eta L + C_{34}\omega \cosh \omega L] = 0 \tag{41}$$

First, we observe that $\bar{P}/EI = -(\eta^2 + \omega^2)$. This can easily be verified through Eqs. (24) and the expressions for γ and P_1:

$$\begin{aligned}
\eta^2 + \omega^2 &= (\rho + ir)^2 + (-\rho + ir)^2 \\
&= 2(\rho^2 - r^2) \\
&= 2\left(\frac{\bar{\beta}}{EI}\right)^{1/2}\left[\frac{1-\gamma}{2} - \frac{1+\gamma}{2}\right] \\
&= -2\left(\frac{\bar{\beta}}{EI}\right)^{1/2}\gamma = -2\left(\frac{\bar{\beta}}{EI}\right)^{1/2}\frac{\bar{P}}{2\sqrt{\bar{\beta}EI}} \\
&= -\frac{\bar{P}}{EI}
\end{aligned}$$

Next, if we add and substract the first two and last two of Eqs. (41), we obtain the following two systems of equations:
Symmetric Buckling:

$$C_{31}\eta^2 \cosh \eta L + C_{32}\omega^2 \cosh \omega L = 0$$
$$C_{31}\omega \sinh \eta L + C_{32}\eta \sinh \omega L = 0$$
(42)

Antisymmetric Buckling:

$$C_{33}\eta^2 \sinh \eta L + C_{34}\omega^2 \sinh \omega L = 0$$
$$C_{33}\omega \cosh \eta L + C_{34}\eta \cosh \omega L = 0$$
(43)

The characteristic equations for both cases must be derived and solved for $\bar{P}_{cr}$.

First, for the symmetric buckling case, the characteristic equation is obtained by requiring the system of Eqs. (42) to have a nontrivial solution:

$$\eta^3 \sinh \omega L \cosh \eta L - \omega^3 \sinh \eta L \cosh \omega L = 0 \qquad (44)$$

Use of Eqs. (24) gives

$$(p + ir)^3[(e^{2irL} - e^{-2irL}) - (e^{2pL} - e^{-2pL})]$$
$$- (p - ir)^3[-(e^{2irL} - e^{-2irL}) - (e^{2pL} - e^{-2pL})] = 0 \qquad (45)$$

Since the second term of the above equation is the complex conjugate of the first term, and since the difference of two complex conjugate pairs is the imaginary part multiplied by $2i$, the characteristic equation is given by

$$I_m[(p + ir)^3(e^{2irL} - e^{-2irL} + e^{-2pL} - e^{2pL})] = 0$$

or

$$I_m\{[p^3 - 3pr^2 + i(3p^2r - r^3)][-2 \sinh 2pL + 2i \sin 2rL]\} = 0$$

This characteristic equation assumes the following final form

$$(3p^2r - r^3) \sinh 2pL = (p^3 - 3pr^2) \sin 2rL \qquad (46)$$

Similarly, for the antisymmetric buckling case, the characteristic equation is

$$(3p^2r - r^3) \sinh 2pL = -(p^3 - 3pr^2) \sin 2rL \qquad (47)$$

Solutions to these equations are presented graphically in Ref. 1.

We see from Eqs. (46) and (47) that, as the length increases, the right-hand side of both equations remains finite. Since $\sinh 2pL$ increases indefinitely, the quantity $3\rho^2 r - r^3$ must approach zero. Therefore, as the length approaches infinity

$$r(3\rho^2 - r^2) = 0$$

Substitution of the expressions for r and ρ from Eq. (24a) gives

$$\gamma = \tfrac{1}{2}$$

and

$$P_{cr} = \sqrt{\beta EI} \qquad (48)$$

An important application of columns on an elastic foundation is in the prediction of wrinkling of the facings in a sandwich construction where the core acts as an elastic foundation (Ref. 2). The mathematical model (buckling equations and boundary conditions) is similar to that used for axisymmetric buckling for thin, circular, cylindrical shells.

PROBLEMS

1. Show that no solution exists (there is no bifurcation point) for $P \geq 2\sqrt{\beta EI}$ when the ends of the column are free (floating ends).

2. Analyze the pinned-pinned column, resting on an elastic foundation, of length $2L$ and uniform flexural stiffness by using the approach outlined in Section 6.4.

3. Find P_{cr} for the clamped-clamped column, resting on an elastic foundation, of length L and uniform flexural stiffness by employing a Rayleigh-Ritz technique and for $0 \leq \beta \leq 49$.

4. Find P_{cr} for the clamped-free column, resting on an elastic foundation, of length L and uniform flexural stiffness by employing the Rayleigh-Ritz technique and for low values of β.

5. Find P_{cr} for a pinned-pinned column, resting on an elastic foundation, of length L and flexural stiffness $EI(x) = EI_1 \sin \pi x/L$ by employing some approximate technique and for low values of β.

REFERENCES

1. HETÉNYI, M., *Beams on Elastic Foundation*, The University of Michigan Press, Ann Arbor, 1946.

2. GOODIER, J. N., and HSU, C. S., "Nonsinusoidal Buckling Modes of Sandwich Plates," *Journal of Aerospace Science*, pp. 525–532, August, 1954.

7

BUCKLING

OF RINGS AND ARCHES

Thin rings and arches (high or low) are often used as structural elements and, when loaded in their plane and in a normal direction, are subject to instability. In this chapter, the analyses of thin circular rings, high circular arches, and low arches are presented. In addition, the analysis of a low half-sine arch, loaded by a half-sine distributed load, resting on an elastic foundation, and pinned at both ends is presented. This is an interesting model because, depending upon the values of the rise parameter and the modulus of the foundation, it exhibits the possibilities of limit point stability (top-of-the-knee buckling), snapthrough buckling through unstable bifurcation, and classical stable bifurcation buckling. For all cases, it is assumed that the behavior of the material is linearly elastic.

7.1 THE THIN CIRCULAR RING

Buckling of thin circular rings was first investigated by Bresse in 1866. Timoshenko and Gere (Ref. 1) present the solution to this problem and a complete historical sketch of other investigations. In addition to the references cited in the text by Timoshenko and Gere, the investigations of Boresi (Ref. 2), Wasserman (Ref. 3), Wempner and Kesti (Ref. 4), and Smith and Simitses (Ref. 5) are important contributions to the solution of this problem.

7.1-1 Kinematic Relations

The geometry and sign convention are given on Fig. 7-1. Let the deformation components of the neutral surface material points be denoted by $w(s)$ and $v(s)$. The strain at any material point z units from the neutral surface is given by

$$\epsilon = \epsilon^0 + z\kappa \tag{1}$$

This equation is based on the assumptions that planes remain plane after deformation, the normals to the neutral axis are inextensional, and the ring is thin (thickness much smaller than the radius). In Eq. (1), ϵ^0 denotes the extensional strain of material points on the neutral surface and κ denotes the change in curvature. We present the development of the expressions of ϵ^0 and κ in terms of the deformation components and their gradients separately.

First, the expression for ϵ^0 is developed. The position vector from the

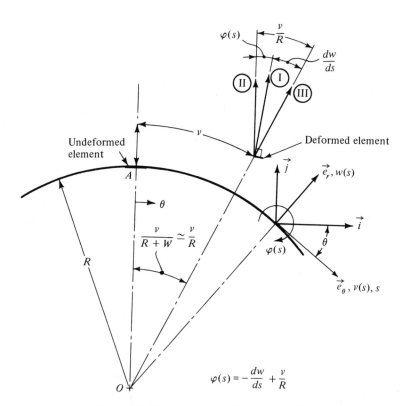

Figure 7-1. Ring geometry and sign convention.

orgin O to any material point on the reference axis is given by

$$\vec{r}(s) = R\vec{e}_r(s) \tag{2}$$

where $\vec{e}_r$ is a unit vector in the radial direction. Let $\vec{d}(s)$ represent the deformation vector; then

$$\vec{d}(s) = w(s)\vec{e}_r + v(s)\vec{e}_\theta \tag{3}$$

The position vector to a material point in the deformed state $\vec{r}^*$ is given by

$$\vec{r}^*(s) = \vec{r}(s) + \vec{d}(s) \tag{4}$$

or

$$\vec{r}^*(s) = R\left(1 + \frac{w}{R}\right)\vec{e}_r + v\vec{e}_\theta \tag{5}$$

Now consider a line segment which is tangent to the s-coordinate line (reference axis) in the undeformed state (of length ds). This segment is given by the vector

$$\vec{dr} = \frac{\vec{dr}}{ds}ds = \vec{e}_\theta\,ds \tag{6}$$

After deformation, the segment is represented by the vector $\vec{dr}^*$, or

$$\vec{dr}^* = \frac{\vec{dr}^*}{ds}ds = \left[\left(\frac{dw}{ds} - \frac{v}{R}\right)\vec{e}_r + \left(1 + \frac{w}{R} + \frac{dv}{ds}\right)\vec{e}_\theta\right]ds \tag{7}$$

The length of this segment is given by

$$|\vec{dr}^*| = \left[\left(\frac{dw}{ds} - \frac{v}{R}\right)^2 + \left(1 + \frac{w}{R} + \frac{dv}{ds}\right)^2\right]^{1/2}ds \tag{8}$$

Now there are two possible definition of extensional strain denoted by ϵ_E (engineering definition) and ϵ_L.

$$\epsilon_E = \frac{|\vec{dr}^*| - |\vec{dr}|}{|\vec{dr}|} \tag{9}$$

$$\epsilon_L = \frac{1}{2}\frac{|\vec{dr}^*|^2 - |\vec{dr}|^2}{|\vec{dr}|^2} \tag{10}$$

From Eq. (9)

$$\frac{|\vec{dr}^*|}{|\vec{dr}|} = \epsilon_E + 1 \tag{11}$$

Substitution of Eq. (11) into Eq. (10) results in

$$\epsilon_L = \epsilon_E + \tfrac{1}{2}\epsilon_E^2 \tag{12}$$

It is clear from Eq. (12) that, for small engineering extensional strains, both definitions give the same results. Therefore, in developing the strain-deformation relations for the thin ring, we will use Eq. (10) to obtain

$$\epsilon^0 = \left(\frac{w}{R} + \frac{dv}{ds}\right) + \frac{1}{2}\left(\frac{dw}{ds} - \frac{v}{R}\right)^2 + \frac{1}{2}\left(\frac{w}{R} + \frac{dv}{ds}\right)^2 \tag{13}$$

Note that the last term is negligibly small by comparison to the first term (in parentheses). Thus,

$$\epsilon^0 = \frac{w}{R} + \frac{dv}{ds} + \frac{1}{2}\left(\frac{dw}{ds} - \frac{v}{R}\right)^2 \tag{14}$$

For small strains and moderately small rotations, the change in curvature can be accurately approximated by (for details see Refs. 5 and 6):

$$\kappa = \frac{d\varphi}{ds} \tag{15}$$

where φ is the rotation of the element from its undeformed state to its deformed state, taken positive as shown in Fig. 7-1.

It is seen from Fig. 7-1, that

$$\varphi = -\frac{dw}{ds} + \frac{v}{R} \tag{16}$$

Therefore

$$\kappa = -\frac{d^2w}{ds^2} + \frac{dv}{R\,ds} \tag{17}$$

Finally, if we use the variable θ instead of $s(\theta = Rs)$, then

$$\epsilon^0 = \frac{w}{R} + \frac{1}{R}\frac{dv}{d\theta} + \frac{1}{2R^2}\left(\frac{dw}{d\theta} - v\right)^2 \tag{18a}$$

$$\kappa = -\frac{1}{R^2}\left(\frac{d^2w}{d\theta^2} - \frac{dv}{d\theta}\right) \tag{18b}$$

and

$$\epsilon = \frac{w}{R} + \frac{1}{R}\frac{dv}{d\theta} + \frac{1}{2R^2}\left(\frac{dw}{d\theta} - v\right)^2 - \frac{z}{R^2}\left(\frac{d^2w}{d\theta^2} - \frac{dv}{d\theta}\right) \tag{19}$$

7.1-2 Equilibrium Equations

Consider the thin circular arch to be loaded by a uniformly destributed load around its circumference with components p_r and p_θ in the radial and tangential directions, respectively. The equilibrium equations for such a configuration are derived using the principle of the stationary value of the total potential

$$\delta U_T = \int_0^{2\pi R} \int_A E(\epsilon^0 + z\kappa)(\delta\epsilon^0 + z\delta\kappa)\, dA\, ds$$

$$- \int_0^{2\pi R} (p_r\delta w + p_\theta\delta v)\, ds = 0 \tag{20}$$

where A is the cross-sectional area of the thin ring. Note that linear elastic behavior is assumed.

Let N and M denote the hoop load and bending moment, respectively,

$$N = \int_A E\epsilon^0 dA = EA\epsilon^0$$

$$M = -\int_A Ez^2\kappa\, dA = -EI\kappa \tag{21}$$

where

$$A = \int_A dA \quad \text{and} \quad I = \int_A z^2\, dA$$

Substitution of Eqs. (21) into Eq. (20) yields

$$\int_0^{2\pi R} [N\delta\epsilon^0 - M\delta\kappa - p_r\delta w - p_\theta\delta v]\, ds = 0 \tag{22}$$

From Eqs. (18), we find the expression for $\delta\epsilon^0$ and $\delta\kappa$:

$$\delta\epsilon^0 = \frac{\delta w}{R} + \frac{1}{R}\frac{d\delta v}{d\theta} + \frac{1}{R^2}\left(\frac{dw}{d\theta} - v\right)\left(\frac{d\delta w}{d\theta} - \delta v\right) \tag{23a}$$

$$\delta\kappa = -\frac{1}{R^2}\left(\frac{d^2\delta w}{d\theta^2} - \frac{d\delta v}{d\theta}\right) \tag{23b}$$

Substitution of Eqs. (23) into Eq. (22), integration by parts, and requiring continuity at any point of the reference axis leads to the following equilibrium equations.

$$-\frac{N}{R} + \frac{d}{ds}\left[N\left(\frac{dw}{ds} - \frac{v}{R}\right)\right] - \frac{d^2M}{ds^2} + p_r = 0$$

$$\frac{dN}{ds} + \frac{N}{R}\left(\frac{dw}{ds} - \frac{v}{R}\right) - \frac{1}{R}\frac{dM}{ds} + p_r = 0 \tag{24}$$

or

$$-NR + \frac{d}{d\theta}\left[N\left(\frac{dw}{d\theta} - v\right)\right] - \frac{d^2M}{d\theta^2} + p_r R^2 = 0$$

$$R\frac{dN}{d\theta} + N\left(\frac{dw}{d\theta} - v\right) - \frac{dM}{d\theta} + p_\theta R^2 = 0 \tag{25}$$

If we assume that the loading is a uniform pressure loading, p, then $p_r = p$, $p_\theta = 0$, and the primary path (prebuckling solution) is characterized by a uniform radial expansion (or contraction). The complete solution of Eqs. (25) for the primary path is given by

$$w^p = \frac{pR^2}{EA}$$

$$v^p = 0 \qquad p_r^p = p$$

$$N^p = pR \tag{26}$$

$$M^p = 0 \qquad p_\theta^p = 0$$

$$\varphi^p = 0$$

7.1-3 Buckling Equations

According to the bifurcation approach, close to the critical load

$$w = w^p + w^* = \frac{p_{cr}R^2}{EA} + w^*$$

$$v = v^p + v^* = v^*$$

$$N = N^p + N^* = p_{cr}R + N^* \tag{27}$$

$$M = M^p + M^* = M^*$$

$$p_r = p_{cr} + p_r^*, \qquad p_\theta = p_\theta^*$$

Substitution of Eqs. (27) into Eqs. (25) yields

$$-N^*R + \frac{d}{d\theta}\left[(p_{cr}R + N^*)\left(\frac{dw^*}{d\theta} - v^*\right)\right] - \frac{d^2M^*}{d\theta^2} + p_r^*R^2 = 0$$

$$R\frac{dN^*}{d\theta} + (p_{cr}R + N^*)\left(\frac{dw^*}{d\theta} - v^*\right) - \frac{dM^*}{d\theta} + p_\theta^*R^2 = 0 \tag{28}$$

Since the starred quantities denote the increments which take the system from the unbuckled state to the adjacent buckled equilibrium state, they can be taken as small (but nonzero) as we wish. Therefore by neglecting N^* as small by comparison to $p_{cr}R$, we have

$$-N^*R + p_{cr}R\left(\frac{d^2w^*}{d\theta^2} - \frac{dv^*}{d\theta}\right) - \frac{d^2M^*}{d\theta^2} + p_r^*R^2 = 0$$

$$R\frac{dN^*}{d\theta} + p_{cr}R\left(\frac{dw^*}{d\theta} - v^*\right) - \frac{dM^*}{d\theta} + p_\theta^*R^2 = 0$$

(29)

Again, because the adjacent state can be taken as close to the primary state as desired, we may use the linearized version of the kinematic relation for the starred quantities:

$$N^* = EA\left(\frac{w^*}{R} + \frac{1}{R}\frac{dv^*}{d\theta}\right)$$

$$M^* = \frac{EI}{R^2}\left(\frac{d^2w^*}{d\theta^2} - \frac{dv^*}{d\theta}\right)$$

(30)

Substitution of Eqs. (30) into Eqs. (29) results in the buckling equations

$$-EA\left(w^* + \frac{dv^*}{d\theta}\right) + p_{cr}R\left(\frac{d^2w^*}{d\theta^2} - \frac{dv^*}{d\theta}\right) - \frac{EI}{R^2}\left(\frac{d^3w^*}{d\theta^4} - \frac{d^3v^*}{d\theta^3}\right)$$
$$+ p_r^*R^2 = 0$$

$$EA\left(\frac{dw^*}{d\theta} + \frac{d^2w^*}{d\theta^2}\right) + p_{cr}R\left(\frac{dw^*}{d\theta} - v^*\right) - \frac{EI}{R^2}\left(\frac{d^3w^*}{d\theta^3} - \frac{d^2v^*}{d\theta^2}\right)$$
$$+ p_\theta^*R^2 = 0$$

(31)

Before we solve Eqs. (31), we must consider the behavior of the load during the buckling process. Since p_r^* and p_θ^* denote the incremental components of the pressure load in the buckled state, the following distinction must be made. There are three possibilities concerning the behavior of the load (cases I, II, and III). In case I, it is assumed that the load remains normal to the deflected reference axis. In Fig. 7-1 the pressure load is in direction I, and its components along the original radial and tangential directions are $p_{cr} \cos \varphi^*$ and $p_{cr} \sin \varphi^*$, respectively. Since φ^* is taken to be small, $p_r^* = 0$ and $p_\theta^* = p_{cr}\varphi^*$. In case II, it is assumed that the load remains parallel to its original direction (direction II in Fig. 7-1). For this case $p_r^* = p_\theta^* = 0$. Finally, in case III, it is assumed that the load is directed toward the initial center of curvature. For this case, $p_r^* = 0$ and $p_\theta^* = p_{cr}(\varphi^* + dw^*/ds)$. In summary

Case I:

$$p_r^* = 0 \qquad p_\theta^* = -\frac{p_{cr}}{R}\left(\frac{dw^*}{d\theta} - v^*\right)$$

Case II:

$$p_r^* = 0 \qquad p_\theta^* = 0$$

(32)

Case III:

$$p_r^* = 0 \qquad p_\theta^* = p_{cr}\frac{v^*}{R}$$

If we now substitute Eqs. (32) into Eqs. (31), we obtain the buckling equation for the three cases of load behavior

$$-EA\left(w^* + \frac{dv^*}{d\theta}\right) + p_{cr}R\left(\frac{d^2w^*}{d\theta^2} - \frac{dv^*}{d\theta}\right) - \frac{EI}{R^2}\left(\frac{d^4w^*}{d\theta^4} - \frac{d^3v^*}{d\theta^3}\right) = 0$$

$$EA\left(\frac{dw^*}{d\theta} + \frac{d^2v^*}{d\theta^2}\right) + p_{cr}R\left(\frac{dw^*}{d\theta} - v^*\right) - \frac{EI}{R^2}\left(\frac{d^3w^*}{d\theta^3} - \frac{d^2v^*}{d\theta^2}\right)$$

$$- p_{cr}R\begin{bmatrix} \dfrac{dw^*}{d\theta} - v^* \\ 0 \\ -v^* \end{bmatrix} = 0$$

Combining like terms in the second equation, we obtain

$$-EA\left(w^* + \frac{dv^*}{d\theta}\right) + p_{cr}R\left(\frac{d^2w^*}{d\theta^2} - \frac{dv^*}{d\theta}\right) - \frac{EI}{R^2}\left(\frac{d^4w^*}{d\theta^4} - \frac{d^3v^*}{d\theta^3}\right) = 0$$

$$EA\left(\frac{dw^*}{d\theta} + \frac{d^2v^*}{d\theta^2}\right) - \frac{EI}{R^2}\left(\frac{d^3w^*}{d\theta^3} - \frac{d^2v^*}{d\theta^2}\right) + p_{cr}R\begin{bmatrix} 0 \\ \dfrac{dw^*}{d\theta} - v^* \\ \dfrac{dw^*}{d\theta} \end{bmatrix} = 0 \qquad (33)$$

We clearly see from Eqs. (33) that the problem has been reduced to an eigenvalue problem in which we seek the smallest value for p_{cr} for which a nontrivial solution exists.

7.1-4 Solution

Before obtaining and discussing the solution, let us multiply both equations by R^2/EI. Then, let $\lambda = p_{cr}R^3/EI$, and $\rho^2 = I/A$, where ρ is the radius of gyration of the cross-sectional geometry

$$-\left(\frac{R}{\rho}\right)^2\left(w^* + \frac{dv^*}{d\theta}\right) + \lambda\left(\frac{d^2w^*}{d\theta^2} - \frac{dv^*}{d\theta}\right) - \left(\frac{d^4w^*}{d\theta^4} - \frac{d^3v^*}{d\theta^3}\right) = 0$$

$$\left(\frac{R}{\rho}\right)^2\left(\frac{dw^*}{d\theta} + \frac{d^2v^*}{d\theta^2}\right) - \left(\frac{d^3w^*}{d\theta^3} - \frac{d^2v^*}{d\theta^2}\right) + \lambda\begin{bmatrix} 0 \\ \dfrac{dw^*}{d\theta} - v^* \\ \dfrac{dw^*}{d\theta} \end{bmatrix} = 0 \qquad (34)$$

Assume solutions of the form

$$w^* = B_n \cos n\theta, \qquad v^* = C_n \sin n\theta$$

or   (35)

$$w^* = B_n \sin n\theta, \qquad v^* = C_n \cos n\theta$$

which satisfy the continuity requirements. Substitution of the first set leads to the following system of homogeneous linear algebraic equations in B_n and C_n:

$$-\left[\left(\frac{R}{\rho}\right)^2 + \lambda n^2 + n^4\right]B_n - \left[\left(\frac{R}{\rho}\right)^2 n + \lambda n + n^3\right]C_n = 0$$

$$-\left[\left(\frac{R}{\rho}\right)^2 n + n^3 + \lambda n \begin{pmatrix} 0 \\ 1 \\ 1 \end{pmatrix}\right]B_n - \left[\left(\frac{R}{\rho}\right)^2 n^2 + n^2 + \lambda \begin{pmatrix} 0 \\ 1 \\ 0 \end{pmatrix}\right]C_n = 0$$

(36)

For a nontrivial solution to exist, the determinant of the coefficients of B_n and C_n must vanish.

The expansion of the determinant yields

$$\lambda^2 n^2 \begin{pmatrix} 0 \\ 0 \\ 1 \end{pmatrix} + \lambda \left(\frac{R}{\rho}\right)^2 \left[\begin{array}{c} n^2(n^2 - 1) \\ (n^2 - 1)^2 \\ n^2(n^2 - 2) - n^4 \left(\frac{\rho}{R}\right)^2 \end{array}\right] + \left(\frac{R}{\rho}\right)^2 n^2(n^2 - 1)^2 = 0$$

(37)

The solutions for λ are:

Case I:

$$\lambda = -(n^2 - 1)$$

Case II:

$$\lambda = -n^2$$ (38)

Case III:

$$\lambda = -\frac{(n^2 - 1)^2}{(n^2 - 2)}$$

To obtain the solution for case III, it is necessary to assume that $(\rho/R)^2 \ll 1$.

The critical value is obtained by minimizing λ with respect to integer values of n. Since $n = 1$ corresponds to rigid body motion (not of interest in this buckling analysis), the critical condition corresponds to $n = 2$.

Case I:

$$\lambda_{cr} = -3, \qquad p_{cr} = -3\frac{EI}{R^3}$$

Case II:

$$\lambda_{cr} = -4, \qquad p_{cr} = -4\frac{EI}{R^3} \qquad (39)$$

Case III:

$$\lambda_{cr} = -4.5, \qquad p_{cr} = -4.5\frac{EI}{R^3}$$

Note that for all three cases, when we assume that $(\rho/R)^2 \ll 1$ and $n = 1$, from the first of Eqs. (36) we have $B_1 = -C_1$.

Next, if we introduce an orthogonal set of unit vectors, $\vec{i}$ and $\vec{j}$ (see Fig. 7-1) in the tangential and radial directions, respectively, when $\theta = 0$, we have

$$\begin{aligned} \vec{e}_\theta &= \cos\theta\,\vec{i} - \sin\theta\,\vec{j} \\ \vec{e}_r &= \sin\theta\,\vec{i} + \cos\theta\,\vec{j} \end{aligned} \qquad (40)$$

The deformation vector, $\vec{d}$, of any material point on the reference axis for $n = 1$ (see Eqs. (35)) is given by

$$\vec{d} = (B_1\cos\theta)\vec{e}_r - (B_1\sin\theta)\vec{e}_\theta \qquad (41)$$

Use of Eqs. (40) and Eq. (41) yields

$$\vec{d} = B_1\vec{j} \qquad (42)$$

Equation (42) shows that, when $n = 1$, we have a rigid body translation.

For all three cases we have found the bifurcation load (classical buckling approach), and no attention is paid to the stability or instability of the system as a rigid body. The reason for considering the three different cases is because all those have been used as models for the real load case, which is pressure. The pressure behavior is best represented by case I. It is difficult to conceive of a true application which is represented by case II. Case III can serve as a mathematical model for the following problem: Consider a thin ring which is loaded by a very large number of closely spaced radial cables pulled together through a stiff, small, rigid ring at the center of the thin ring. Singer and Babcock (Ref. 7) have shown that for case II the thin ring is unstable as a rigid body and will rotate under arbitrarily small pressure.

7.2   HIGH CIRCULAR ARCHES UNDER PRESSURE

The buckling of a high circular arch under uniform pressure has been presented in Ref. 1. As discussed in Ref. 1, the solution was first obtained by Hurlbrink and the problem was also investigated by Timoshenko and Nicolai. The solution presented herein is based on the use of Eqs. (34). In order to apply these equations, it is assumed that first the arch is uniformly contracted (see Fig. 7-2). On this basis a primary state exists, which is identical to that of a complete circular ring. Then at the instant of buckling, the supports become immovable and the arch buckles as shown in Fig. 7-2.

If the arch is simply supported at both ends, the boundary conditions at $\theta = \pm\alpha$, for the incremental quantities, are given by

$$w^* = 0$$

$$M^* = \frac{EI}{R^2}\left(\frac{d^2w^*}{d\theta^2} - \frac{dv^*}{d\theta}\right) = 0 \tag{43}$$

$$N^* = \frac{EA}{R}\left(w^* + \frac{dv^*}{d\theta}\right) = 0$$

These boundary conditions are satisfied by the following assumed form of solution:

$$w^* = B_n \sin \frac{n\pi\theta}{\alpha}$$

$$v^* = C_n \cos \frac{n\pi\theta}{\alpha} \tag{44}$$

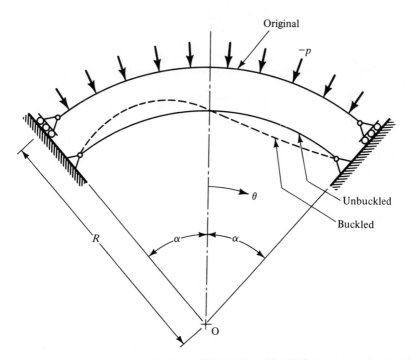

Figure 7-2. Geometry of pinned circular arch.

Substitution of Eqs. (44) into Eqs. (43) gives

$$-\left[\left(\frac{R}{\rho}\right)^2 + \lambda\left(\frac{n\pi}{\alpha}\right)^2 + \left(\frac{n\pi}{\alpha}\right)^4\right]B_n$$

$$+\left[\left(\frac{R}{\rho}\right)\left(\frac{n\pi}{\alpha}\right) + \lambda\left(\frac{n\pi}{\alpha}\right) + \left(\frac{n\pi}{\alpha}\right)^3\right]C_n = 0$$

$$\left[\left(\frac{R}{\rho}\right)^2\left(\frac{n\pi}{\alpha}\right) + \left(\frac{n\pi}{\alpha}\right)^3 + \lambda\left(\frac{n\pi}{\alpha}\right)\begin{pmatrix}0\\1\\1\end{pmatrix}\right]B_n \tag{45}$$

$$-\left[\left(\frac{R}{\rho}\right)^2\left(\frac{n\pi}{\alpha}\right)^2 + \left(\frac{n\pi}{\alpha}\right)^2 + \lambda\begin{pmatrix}0\\1\\0\end{pmatrix}\right]C_n = 0$$

For a nontrivial solution to exist, the determinant of the coefficients of B_n and C_n must vanish. The expansion of the determinant yields an expression similar to Eq. (37), except that whenever n appears in Eq. (37), we must put $(n\pi/\alpha)$. The solutions for λ are:

Case I:

$$\lambda = -\left[\left(\frac{n\pi}{\alpha}\right)^2 - 1\right]$$

Case II:

$$\lambda = -\left(\frac{n\pi}{\alpha}\right)^2 \tag{46}$$

Case III:

$$\lambda = -\frac{[(n\pi/\alpha)^2 - 1]^2}{[(n\pi/\alpha)^2 - 2]}$$

The critical condition corresponds to the smallest n, which is one for this case. Therefore,

Case I:

$$p_{cr} = -\frac{EI}{R^3}\left[\left(\frac{\pi}{\alpha}\right)^2 - 1\right]$$

Case II:

$$p_{cr} = -\frac{EI}{R^3}\left(\frac{\pi}{\alpha}\right)^2 \tag{47}$$

Case III:

$$p_{cr} = -\frac{EI}{R^3}\frac{[(\pi/\alpha)^2 - 1]^2}{(\pi/\alpha)^2 - 2}$$

The solution for case I is the same as the one reported in Ref. 1. For this case we note that, when $\alpha = \pi$, we have a complete ring and $p_{cr} = 0$. The reason for this undesirable result is that we have a complete ring with a hinge, and it is free to rotate as a rigid body about this hinge for arbitrarily small pressure. The continuous complete ring corresponds to $\alpha = \pi/2$. When $\alpha = \pi/2$, Eqs. (47) are identical to Eqs. (39).

These results, which are derived by assuming the buckling mode of Fig. 7-2, are not applicable to shallow arches. The low arch is treated in a later section.

The solution to the clamped arch is presented in Ref. 1. This solution is due to E. L. Nicolai:

$$p_{cr} = -\frac{EI}{R^3}(k^2 - 1) \tag{48}$$

where k is the solution of the following transcendental equation

$$k \tan \alpha \cot k\alpha = 1 \tag{49}$$

7.3 ALTERNATE APPROACH FOR RINGS AND ARCHES

An alternate approach for solving the ring and high arch problem, is to eliminate v^* from Eqs. (34) and obtain a single higher-order buckling equation in w^* alone. This single equation is then solved, subject to the appropriate boundary conditions (when applicable). This approach is used in this discussion for load case I. Equations (34) may be written as

$$
\begin{aligned}
L_1 w^* + L_2 v^* &= 0 \\
L_3 w^* + L_4 v^* &= 0
\end{aligned}
\tag{50}
$$

where L_i ($i = 1, 2, 3,$ and 4) are the following differential operators

$$
\begin{aligned}
L_1 &= -\left(\frac{R}{\rho}\right)^2 + \lambda \frac{d^2}{d\theta^2} - \frac{d^4}{d\theta^4} \\
L_2 &= -\left[\lambda + \left(\frac{R}{\rho}\right)^2\right] \frac{d}{d\theta} + \frac{d^3}{d\theta^3} \\
L_3 &= \left(\frac{R}{\rho}\right)^2 \frac{d}{d\theta} - \frac{d^3}{d\theta^3} \\
L_4 &= \left[1 + \left(\frac{R}{\rho}\right)^2\right] \frac{d^2}{d\theta^2}
\end{aligned}
\tag{51}
$$

Since these operators are linear, they are commutative

$$L_1 L_4 = L_4 L_1, \qquad L_2 L_4 = L_4 L_2, \qquad \text{etc.}$$

By operating with L_4 on the first of Eqs. (50), with L_2 on the second of Eqs. (50), and by subtracting the two resulting equations, we have

$$(L_1 L_4 - L_3 L_2)w^* = 0 \tag{52}$$

Substitution of the expressions, Eqs. (51), for the operators yields the following single buckling equation:

$$\frac{d^6w^*}{d\theta^6} + (2 - \lambda)\frac{d^4w^*}{d\theta^4} + (1 - \lambda)\frac{d^2w^*}{d\theta^2} = 0 \tag{53}$$

If we let the solution to Eq. (53) be of the form $e^{\gamma\theta}$, we obtain

$$\gamma^2[\gamma^4 - (\lambda - 2)\gamma^2 - (\lambda - 1)] = 0 \tag{54}$$

and the six roots are

$$\gamma_1 = +i\sqrt{1 - \lambda} = ik$$
$$\gamma_2 = -i\sqrt{1 - \lambda} = -ik$$
$$\gamma_3 = +i \tag{55}$$
$$\gamma_4 = -i$$
$$\gamma_5 = \gamma_6 = 0$$

where

$$k = \sqrt{1 - \lambda} \quad \text{and} \quad \lambda = 1 - k^2$$

From this, the general solution to Eq. (53) is

$$w^* = A_1 \sin k\theta + A_2 \cos k\theta + A_3 \sin \theta + A_4 \cos \theta + A_5\theta + A_6 \tag{56}$$

Note that λ is a negative number because buckling of the ring is possible only when the uniform radial pressure is compressive and k is the positive square root. Similarly, if we eliminate w^* from Eqs. (50), we obtain a single higher-order equation in v^* [same as Eq. (53)]. The solution for v^* is

$$v^* = B_1 \sin k\theta + B_2 \cos k\theta + B_3 \sin \theta + B_4 \cos \theta + B_5\theta + A_7 \tag{57}$$

If we substitute Eqs. (56) and (57) into Eqs. (50), we have

$$B_1 = -\frac{A_2}{k}\frac{k^2 + (R/\rho)^2}{1 + (R/\rho)^2}, \quad B_2 = \frac{A_1}{k}\frac{k^2 + (R/\rho)^2}{1 + (R/\rho)^2}$$

$$B_3 = -A_4, \quad B_4 = A_3, \quad A_5 \equiv 0 \tag{58}$$

$$B_5 = -\frac{(R/\rho)^2}{1 - k^2 + (R/\rho)^2}$$

Next, if we make use of the thin ring assumption, $(\rho/R)^2 \ll 1$, and if we substitute Eqs. (58) into Eq. (57), we have

$$v^* = \frac{A_1}{k}\cos k\theta - \frac{A_2}{k}\sin k\theta + A_3 \cos \theta - A_4 \sin \theta - A_6\theta + A_7 \tag{59}$$

Substitution of Eqs. (56) and (59) into Eqs. (30) yields the following results for the incremental hoop load, N^*, and the incremental bending moment, M^*,

$$N^* \equiv 0$$

$$M^* = \frac{EI}{R^2}[(1 - k^2)(A_1 \sin k\theta + A_2 \cos k\theta) + A_6] \tag{60}$$

We obtain the expression for the rotation at any point by substituting Eqs. (56) and (59) into Eq. (16):

$$\varphi^* = -\frac{1}{R}\left[\frac{k^2 - d}{k}(A_1 \cos k\theta - A_2 \sin k\theta) + A_6\theta - A_7\right] \tag{61}$$

Finally, the expression for the radial shear, Q_r^*, is given by

$$Q_r^* = -\frac{1}{R}\frac{dM^*}{d\theta} = \frac{EJ}{R^3}k(k^2 - 1)(A_1 \cos k\theta - A_2 \sin k\theta) \tag{62}$$

7.3-1 The Circular Ring

For this particular case, the characteristic equation is obtained from requiring continuity in w^*, v^*, φ^*, M^*, and Q_r^* at the ring reference axis. The continuity equations are:

$$w^*(0) = w^*(2\pi)$$
$$v^*(0) = v^*(2\pi)$$
$$\varphi^*(0) = \varphi^*(2\pi) \tag{63}$$
$$M^*(0) = M^*(2\pi)$$
$$Q_r^*(0) = Q_r^*(2\pi)$$

Substitution of Eqs. (56), and Eqs. (59) through (62) into Eqs. (63) gives

$$A_1 \sin 2k\pi + A_2 (\cos 2k\pi - 1) = 0 \tag{64}$$

$$A_1\frac{1}{k}(\cos 2k\pi - 1) - A_2\frac{1}{k}\sin 2k\pi - A_6(2\pi) = 0 \tag{65}$$

$$A_1(1 - k^2)\sin 2k\pi + A_2(1 - k^2)(\cos 2k\pi - 1) + A_6 = 0 \tag{66}$$

$$\frac{k^2 - 1}{k}[A_1(\cos 2k\pi - 1) - A_2 \sin 2k\pi] + A_6(2\pi) - A_7 = 0 \tag{67}$$

and

$$A_1(\cos 2k\pi - 1) - A_2 \sin 2k\pi = 0 \tag{68}$$

The first four of these equations comprise a system of four linear homogeneous algebraic equations in A_1, A_2, A_6, and A_7. For a nontrivial solution to exist, the determinant of the coefficients must vanish.

$$\begin{vmatrix} \sin 2k\pi & \cos 2k\pi - 1 & 0 & 0 \\ \dfrac{1}{k}(\cos 2k\pi - 1) & -\dfrac{1}{k}\sin 2k\pi & 2\pi & 0 \\ (1 - k^2)\sin 2k\pi & (1 - k^2)(\cos 2k\pi - 1) & 1 & 0 \\ \dfrac{k^2 - 1}{k}(\cos 2k\pi - 1) & -\dfrac{k^2 - 1}{k}\sin 2k\pi & 2\pi & -1 \end{vmatrix} = 0 \qquad (69)$$

The expansion of this determinant yields

$$\cos 2k\pi = 1 \qquad (70)$$

Equation (70) is the characteristic equation, and the solution is

$$2k\pi = 2n\pi, \qquad n = 0, 1, 2, \dots \qquad (71)$$

From Eq. (71), the critical load parameter, λ, corresponds to $n = 2$ and $\lambda_{cr} = -3$.

Since $k_{cr} = n = 2$, then from Eqs. (64) through (67),

$$A_6 = A_7 = 0 \qquad (72)$$

In addition, for this value of k, Eq. (68) is satisfied and thus continuity in shear Q_r^* does exist. Note that A_3 and A_4 do not appear in any of the continuity equations. This is not surprising because the A_3 and A_4 terms, in the expressions for w^* and v^*, denote rigid body translation.

7.3-2 The Pinned Circular Arch

For this particular case we assume that the ring supports are on rollers and a membrane state exists (see Fig. 7-2). At the instant of buckling, the pin supports become immovable. Thus the boundary conditions for the buckling equations, Eqs. (34), are

$$w^*(-\alpha) = w^*(\alpha) = 0$$
$$M^*(-\alpha) = M^*(\alpha) = 0 \qquad (73)$$
$$v^*(-\alpha) = v^*(\alpha) = 0$$

Using Eqs. (56), (59), and (60) in Eqs. (73), we obtain

$$\mp A_1 \sin k\alpha + A_2 \cos k\alpha \mp A_3 \sin \alpha + A_4 \cos \alpha + A_6 = 0 \qquad (74)$$
$$(1 - k^2)(\mp A_1 \sin k\alpha + A_2 \cos k\alpha) + A_6 = 0 \qquad (75)$$
$$A_1 \frac{1}{k}\cos k\alpha \pm A_2 \frac{1}{k}\sin k\alpha + A_3 \cos \alpha \pm A_4 \sin \alpha \pm A_6\alpha + A_7 = 0 \qquad (76)$$

Addition and subtraction of each pair of equations, Eqs. (74), (75), and (76), yield the following two sets of linear homogeneous algebraic equations in A_1, A_3, A_7, and A_2, A_4, and A_6.

$$A_1 \sin k\alpha + A_3 \sin \alpha = 0$$
$$A_1(1 - k^2) \sin k\alpha = 0 \tag{77}$$
$$A_1 \frac{1}{k} \cos k\alpha + A_3 \cos \alpha + A_7 = 0$$

$$A_2 \cos k\alpha + A_4 \cos \alpha + A_6 = 0$$
$$A_2(1 - k^2) \cos k\alpha + A_6 = 0 \tag{78}$$
$$A_2 \frac{1}{k} \sin k\alpha + A_4 \sin \alpha + A_6\alpha = 0$$

Equations (77) correspond to an antisymmetric mode of deformation, w^*, while Eqs. (78) correspond to a symmetric mode, with respect to $\theta = 0$.

Antisymmetric Buckling. From Eqs. (77), we obtain the characteristic equation

$$\begin{vmatrix} \sin k\alpha & \sin \alpha & 0 \\ (1 - k^2) \sin k\alpha & 0 & 0 \\ \frac{1}{k} \cos k\alpha & \cos \alpha & 1 \end{vmatrix} = 0 \tag{79}$$

The expansion of the determinant gives

$$\sin k\alpha = 0 \tag{80}$$

and

$$k\alpha = n\pi, \quad n = 1, 2, 3, \ldots \tag{81}$$

Thus $k_{cr} = \pi/\alpha$ and $\lambda_{cr} = -(\pi/\alpha)^2 + 1$, as expected. Note that $A_3 = A_7 = 0$ for this case.

Symmetric Buckling. The characteristic equation for symmetric buckling is obtained from Eqs. (78):

$$\begin{vmatrix} \cos k\alpha & \cos \alpha & 1 \\ (1 - k^2) \cos k\alpha & 0 & 1 \\ \frac{1}{k} \sin k\alpha & \sin \alpha & \alpha \end{vmatrix} = 0 \tag{82}$$

The expansion of the determinant yields

$$\tan k\alpha = (k\alpha)^3 \frac{\tan \alpha - \alpha}{\alpha^3} + (k\alpha) \tag{83}$$

The corresponding expression for w^* is given by

$$w^* = A_2\left[\cos k\theta - k^2 \frac{\cos k\alpha}{\cos \alpha} \cos \theta + (k^2 - 1) \cos k\alpha\right] \qquad (84)$$

Careful study of Eqs. (83) and (84) reveals that the first acceptable root of Eq. (83) for symmetric buckling is near $3\pi/2$, or $|\lambda_{cr}|$ for symmetric buckling is greater than $|\lambda_{cr}|$ for antisymmetric buckling.

7.3-3 The Clamped Arch

Following the same line of thinking as in the pinned arch case, we find that the boundary conditions for this case are

$$w^*(-\alpha) = w^*(\alpha) = 0$$
$$\varphi^*(-\alpha) = \varphi^*(\alpha) = 0 \qquad (85)$$
$$v^*(-\alpha) = v^*(\alpha) = 0$$

Use of Eqs. (56), (59), and (61) in Eqs. (85) yields

$$\mp A_1 \sin k\alpha + A_2 \cos k\alpha \mp A_3 \sin \alpha + A_4 \cos \alpha + A_6 = 0 \qquad (86)$$

$$\frac{k^2 - 1}{k}(A_1 \cos k\alpha \mp A_2 \sin k\alpha) \pm A_6 \alpha - A_7 = 0 \qquad (87)$$

$$A_1\frac{1}{k}\cos k\alpha \pm A_2\frac{1}{k}\sin k\alpha + A_3 \cos \alpha \pm A_4 \sin \alpha \pm A_6\alpha + A_7 = 0 \qquad (88)$$

Following the same steps as in the case of the pinned arch we have:

Antisymmetric Buckling.

$$A_1 \sin k\alpha + A_3 \sin \alpha = 0$$
$$A_1\frac{k^2 - 1}{k} \cos k\alpha - A_7 = 0 \qquad (89)$$
$$A_1\frac{1}{k} \cos k\alpha + A_3 \cos \alpha + A_7 = 0$$

From Eqs. (89) the characteristic equation is

$$\begin{vmatrix} \sin k\alpha & \sin \alpha & 0 \\ \dfrac{k^2 - 1}{k}\cos k\alpha & 0 & -1 \\ \dfrac{1}{k}\cos k\alpha & \cos \alpha & 1 \end{vmatrix} = 0 \qquad (90)$$

The expansion of the determinants yields

$$k \tan \alpha \cot k\alpha = 1 \tag{91}$$

This equation is identical with Eq. (49).

Symmetric Buckling.

$$A_2 \cos k\alpha + A_4 \cos \alpha + A_6 = 0$$

$$A_2 \frac{k^2 - 1}{k} \sin k\alpha - A_6 \alpha = 0 \tag{92}$$

$$A_2 \frac{1}{k} \sin k\alpha + A_4 \sin \alpha + A_6 \alpha = 0$$

The characteristic equation is

$$\begin{vmatrix} \cos k\alpha & \cos \alpha & 1 \\ \dfrac{k^2 - 1}{k} \sin k\alpha & 0 & -\alpha \\ \dfrac{1}{k} \sin k\alpha & \sin \alpha & \alpha \end{vmatrix} = 0 \tag{93}$$

which yields

$$\cot k\alpha = \frac{(k\alpha)}{\alpha} \left(\frac{1}{\tan \alpha} - \frac{1}{\alpha} \right) + \frac{1}{(k\alpha)} \tag{94}$$

Note that, when $\alpha = \pi$, both Eqs. (91) and (94) yield $k_{cr} = 2$ and therefore $\lambda_{cr} = -3$.

7.4 SHALLOW ARCHES

Shallow arches have been used widely as structural elements. One important response of such elements when loaded transversely (see Fig. 7-3) is snap-through buckling or oil-canning. This buckling phenomenon is characterized by a visible and sudden jump from one equilibrium configuration to another for which displacements are distinctly larger than the first. The significance of snapthrough buckling, insofar as it illustrates certain important features in more complicated buckling problems of plates and shells, was pointed out by Marguerre (Ref. 8), who constructed a simplified mechanical model to demonstrate these features. Timoshenko (Ref. 9) obtained an approximate solution to the problem of a low arch under a uniformly distributed transverse load. Biezeno (Ref. 10) considered the problem of a low arch loaded transversely at the midpoint with a concentrated load. According to Ref. 1, this problem was first discussed by Navier (Ref. 11). Fung and Kaplan (Ref. 12) investigated the problem of pinned low arches of various initial shapes

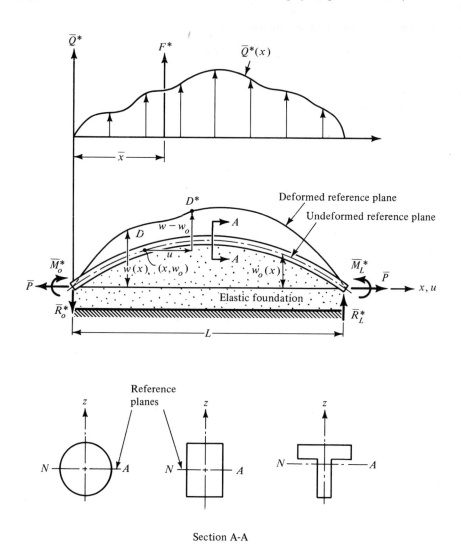

Section A-A

Figure 7-3. The shallow arch: geometry and sign convention.

(parabolic, half-sine, circular, etc.), and spatial distributions of the lateral load. They also considered the effect of prestress on the critical value of the load. About the same time, Hoff and Bruce (Ref. 13) presented results for the pinned half-sine low arch under a half-sine distributed load, as a special case of their dynamic analysis of the buckling of laterally loaded low arches. The results of these two analyses show that a very shallow arch snaps through symmetrically (limit point instability), whereas a higher shallow arch snaps through asymmetrically (unstable bifurcation).

Gjelsvik and Bodner (Ref. 14) obtained an approximate solution to the problem of a clamped low circular arch with a concentrated lateral load at the midpoint. Schreyer and Masur (Ref. 15) obtained an exact solution to this problem (and to the case of uniform pressure), and they showed that for the concentrated load case, the arch snaps-through symmetrically regardless of the value of the rise parameter. Masur and Lo (Ref. 16) presented a general discussion of the behavior of the shallow circular arch regarding buckling, post-buckling and imperfection sensitivity. The effects of inelastic material behavior have been considered by Franciosi, Augusti and Sparacio (Ref. 17), by Onat and Shu (Ref. 18), and by Lee and Murphy (Ref. 19). Experimental results have been reported in Ref. 12, and by Roorda (Ref. 20). Finally, snapping of low pinned arches resting on an elastic foundation has been investigated by Simitses (Ref. 21). This work is presented in a later section because it provides an interesting model for stability studies. This model exhibits all forms of experimentally observed buckling phenomena.

7.4-1 Mathematical Formulation

The equilibrium equations and proper boundary conditions will be derived first. Consider a slender arch of small initial curvature. We assume that all of the assumptions of slender beams are satisfied except that we now have initial shape, such that material points on the undeformed midline (midplane) are characterized by $w_0(x)$ (see Fig. 7-3). Let $u(x)$ and $w(x)$ denote the location of material points on the deformed midline. On the basis of these assumptions, the strain at any material point is given by

$$\epsilon = \epsilon^0 + z\kappa \tag{95}$$

where ϵ^0 and κ denote the reference-plane extensional strain and change in curvature, respectively. The appropriate kinematic relations are derived by reference to Fig. 7-3. Let D and D^* denote the undeformed and deformed positions of a material point of the reference line [intersection of the plane of symmetry, xz, with the reference plane (neutral surface)]. The coordinates of D and D^* are (x, w_0) and $[x + u(x), w]$, respectively. If ds and $(ds)^*$ denote the undeformed and deformed lengths of elements on the reference line, then for small strains, the reference plane extensional strain, ϵ^0, is given by

$$\epsilon^0 = \frac{1}{2} \frac{(ds^*)^2 - (ds)^2}{(ds)^2} \tag{96}$$

Since

$$(ds^*) = (dx + du)^2 + (dw)^2$$
$$(ds)^2 = (dx)^2 + (dw_0)^2 \tag{97}$$

then

$$\epsilon^0 = \frac{du}{dx} + \frac{1}{2}\left(\frac{dw}{dx}\right)^2 - \frac{1}{2}\left(\frac{dw_0}{dx}\right)^2 \tag{98}$$

The expression in Eq. (98) is based on the assumption that $(dw_0/dx)^2 \ll 1$ and $du/dx \ll 1$.

In addition, for small initial curvature and for $(dw/dx)^2 \ll 1$, the expression for the change in curvature, κ, is given by

$$\kappa = \frac{d\varphi}{dx} = -\left(\frac{d^2w}{dx^2} - \frac{d^2w_0}{dx^2}\right) \tag{99}$$

Assuming that the behavior of the material is linearly elastic and denoting by P and M the axial force and bending moment, respectively, we have

$$P = EA\epsilon^0$$
$$M = EI\kappa \tag{100}$$

where A is the cross-sectional area and I is the second moment of this area about the neutral axis.

We obtain the equilibrium equations by making use of the principle of the stationary value of the total potential. The total potential consists of the sum of the energy stored in the system and the potential of the external forces. The energy stored in the system is the sum of the stretching strain energy, bending strain energy, and energy stored into the foundation. If we let U_m^*, U_b^*, and U_f^* denote the linear stretching, bending, and foundation energy densities, we can write the following expressions

$$U_m^* = \frac{1}{2}EA(\epsilon^0)^2 = \frac{P^2}{2EA}$$

$$U_b^* = \frac{1}{2}EI\kappa^2 = \frac{M^{*2}}{2EI} \tag{101}$$

$$U_f^* = \frac{1}{2}\bar{\beta}(w - w_0)^2$$

where $\bar{\beta}$ is the modulus of the foundation (same as in Chapter 6). The potential of the external forces, U_{Pr}^*, includes the contributions of the distributed load $Q^*(x)$, the concentrated loads F^*, $\bar{R}^*$, and $\bar{P}$, and the couples $\bar{M}^*$.

$$U_{Pr}^* = -\int_0^L [Q^* + F^*\delta^*(x - \bar{x})](w - w_0)\,dx$$
$$- [\bar{R}^*(w - w_0)]_0^L +]\bar{M}^*\varphi]_0^L - [\bar{P}u]_0^L \tag{102}$$

where $\delta^*(x - \bar{x})$ is the Dirac-delta function.

Since Eqs. (101) denote linear densities, the total potential is obtained through integration of these expressions, over the entire length, and added to U_{Pr}^*, Eq. (102).

$$U_T^* = \int_0^L \left[\frac{P^2}{2EA} + \frac{M^{*2}}{2EI} + \frac{1}{2}\bar{\beta}(w - w_0)^2 - \{Q^* + F^*\delta^*(x - \bar{x})\}(w - w_0) \right] dx$$
$$- [\bar{R}^*(w - w_0)]_0^L + [\bar{M}^*\varphi]_0^L - [\bar{P}u]_0^L \tag{103}$$

Before we proceed with the derivation of the equilibrium equations and the proper boundary conditions, it is convenient to express all of the parameters in a nondimensional form. We do this by introducing the following parameters:

$$\xi = \frac{\pi x}{L}, \qquad \eta(\xi) = \frac{w(x)}{\rho}, \qquad v(\xi) = \frac{u(x)}{\rho}, \qquad \delta^*(x - \bar{x}) = \rho\delta(\xi - \bar{\xi})$$

$$q(\xi) = \frac{\rho Q^*(x)}{P_E \bar{\epsilon}_E}, \qquad \beta = \frac{\bar{\beta}L^4}{\pi^4 EI}, \qquad F = \frac{F^*}{P_E \bar{\epsilon}_E} \tag{104}$$

$$\bar{R} = \frac{\bar{R}^*}{P_E \bar{\epsilon}_E}, \qquad p = \frac{P}{P_E}, \qquad M = \frac{M^*}{\rho P_E}, \qquad U_T = \frac{4U_T^*}{P_E \bar{\epsilon}_E L}, \qquad (\ )' = \frac{d}{d\xi}$$

where

$$\rho^2 = \frac{I}{A}, \qquad P_E = \frac{\pi^2 EI}{L^2}, \quad \text{and} \quad \bar{\epsilon} = \left(\frac{\pi\rho}{L}\right)^2$$

Note that ρ is the radius of gyration for the cross-sectional area, P_E is the Euler load for a pinned column of length L, and $\bar{\epsilon}_E$ is the corresponding Euler strain.

With these new nondimensionalized parameters, Eqs. (100), (101), and (102) become

$$P = \frac{P_E}{2}\left[2\frac{v'}{\bar{\epsilon}_E^{1/2}} + (\eta')^2 - (\eta_0')^2 \right] \tag{105}$$

$$M^* = -\rho P_E(\eta'' - \eta_0'')$$

$$U_m^* = \frac{P_E \bar{\epsilon}_E}{8}\left[2\frac{v'}{\bar{\epsilon}_E^{1/2}} + (\eta')^2 - (\eta_0')^2 \right]$$

$$U_b^* = \frac{P_E \bar{\epsilon}_E}{2}(\eta'' - \eta_0'')^2 \tag{106}$$

$$U_f^* = \frac{P_E \bar{\epsilon}_E}{2}\beta(\eta - \eta_0)^2$$

$$U_{PT} = -\frac{1}{\pi}\int_0^\pi [4q + 4F\delta(\xi - \bar{\xi})](\eta - \eta_0)\,d\xi - \frac{1}{\pi}[\bar{R}(\eta - \eta_0)]_0^\pi(4\bar{\epsilon}_E^{1/2})$$
$$- \frac{4}{\pi}[\bar{M}(\eta' - \eta_0')]_0^\pi - \frac{1}{\pi}\left(\frac{4}{\bar{\epsilon}_E^{1/2}}\right)[\bar{p}v]_0^\pi \tag{107}$$

Finally, the nondimensionalized total potential, Eq. (103), becomes

$$U_T = \frac{4U_T^*}{P_E \bar{\epsilon}_E L} = \frac{1}{\pi} \int_0^\pi \left[\frac{1}{2} \left\{ 2\frac{v'}{\bar{\epsilon}_E^{1/2}} + (\eta')^2 - (\eta_0')^2 \right\}^2 + 2(\eta'' - \eta_0'')^2 \right.$$

$$+ 2\beta(\eta - \eta_0)^2 - 4q(\eta - \eta_0) - 4F\delta(\xi - \bar{\xi})(\eta - \eta_0) \Big] d\xi \qquad (108)$$

$$- \frac{1}{\pi} \left([\bar{R}(\eta - \eta_0)]_0^\pi (4\bar{\epsilon}_E^{1/2}) + 4[\bar{M}(\eta' - \eta_0')]_0^\pi + \frac{4}{\bar{\epsilon}_E^{1/2}}[\bar{p}v]_0^\pi \right)$$

According to the principle of the stationary value of the total potential, the first variation of the total potential must be zero for equilibrium. To accomplish this, we express the functional U_T in terms of $v(\xi) + \epsilon_1\zeta(\xi)$ and $\eta(\xi) + \epsilon_2\gamma(\xi)$, where $\zeta(\xi)$ and $\gamma(\xi)$ are admissible functions of ξ, and ϵ_1 and ϵ_2 are small arbitrary constants. Thus $\epsilon_1\zeta$ and $\epsilon_2\gamma$ denote the variations in v and η, respectively (δv and $\delta\eta$).

$$U_T[v + \epsilon_1\zeta, \eta + \epsilon_2\gamma] = \frac{1}{\pi} \int_0^\pi \left\{ \frac{1}{2} \left[2\frac{v' + \epsilon_1\zeta'}{\bar{\epsilon}_E^{1/2}} + (\eta' + \epsilon_2\gamma')^2 - (\eta_0')^2 \right]^2 \right.$$

$$+ 2(\eta'' + \epsilon_2\gamma'' - \eta_0'')^2 + 2\beta(\eta + \epsilon_2\gamma - \eta_0)^2 - 4q(\eta + \epsilon_2\gamma - \eta_0)$$

$$- 4F\delta(\xi - \bar{\xi})(\eta + \epsilon_2\gamma - \eta_0) \Big\} d\xi - \frac{1}{\pi}(4\bar{\epsilon}_E^{1/2})[\bar{R}(\eta + \epsilon_2\gamma - \eta_0)]_0^\pi$$

$$- \frac{4}{\pi}[\bar{M}(\eta' + \epsilon_2\gamma' - \eta_0')]_0^\pi - \frac{1}{\pi}\left(\frac{4}{\bar{\epsilon}_E^{1/2}}\right)[\bar{p}(v + \epsilon_1\zeta)]_0^\pi \qquad (109)$$

If we perform the operations indicated in the integrand and group terms according to the powers in ϵ_1 and ϵ_2, we recognize that the terms that do not contain ϵ's denote $U_T[v, \eta]$ and thus

$$U_T[v + \epsilon_1\zeta, \eta + \epsilon_2\gamma] = U_T[v, \eta] + \frac{1}{\pi}\left\{ \frac{\epsilon_1}{\bar{\epsilon}_E^{1/2}} \int_0^\pi 2 \left[2\frac{v'}{\bar{\epsilon}_E^{1/2}} + (\eta')^2 - (\eta_0')^2 \right]\zeta' d\xi \right.$$

$$- \frac{\epsilon_1}{\bar{\epsilon}_E^{1/2}}[4\bar{p}\zeta]_0^\pi + \epsilon_2 \int_0^\pi \left(2\left[\frac{2v'}{\bar{\epsilon}_E^{1/2}} + (\eta')^2 - (\eta_0')^2 \right]\eta'\gamma' + 4(\eta'' - \eta_0'')\gamma'' \right.$$

$$+ 4\beta(\eta - \eta_0)\gamma - 4q\gamma - 4F\delta(\xi - \bar{\xi})\gamma \Big) d\xi - \epsilon_2[4\bar{\epsilon}_E^{1/2}\bar{R}\gamma]_0^\pi - \epsilon_2[4\bar{M}\gamma']_0^\pi \Big\}$$

$$+ \frac{1}{\pi}\left\{ \epsilon_1^2 \int_0^\pi \frac{2}{\bar{\epsilon}_E}(\zeta')^2 d\xi + \epsilon_1\epsilon_2 \int_0^\pi \frac{4}{\bar{\epsilon}_E^{1/2}}\eta'\zeta'\gamma' d\xi \right.$$

$$+ \epsilon_2^2 \int_0^\pi \left[2(\eta')^2(\gamma')^2 + \left\{ \frac{2v'}{\bar{\epsilon}_E^{1/2}} + (\eta')^2 - (\eta_0')^2 \right\}(\gamma')^2 + 2(\gamma'')^2 + 2\beta\gamma^2 \right] d\xi \Big\}$$

$$+ \frac{1}{\pi}\left[\frac{\epsilon_1\epsilon_2^2}{\bar{\epsilon}_E^{1/2}} \int_0^\pi 2\zeta'(\gamma')^2 d\xi + 2\epsilon_2^3 \int_0^\pi (\eta')(\gamma')^3 d\xi \right] + \frac{\epsilon_2^4}{2\pi} \int_0^\pi (\gamma')^4 d\xi \qquad (110)$$

From Eqs. (105), the first variation becomes

$$\delta^1 U_T = \frac{1}{\pi} \left\{ \frac{4\epsilon_1}{\bar{\epsilon}_E^{1/2}} \left[\int_0^\pi p\zeta' \, d\xi - (\bar{p}\zeta) |_0^\pi \right] \right.$$

$$+ 4\epsilon_2 \left(\int_0^\pi [p\eta'\gamma' + (\eta'' - \eta_0'')\gamma'' + \beta(\eta - \eta_0)\gamma - q\gamma \right.$$

$$\left. \left. - F\delta(\xi - \bar{\xi})\gamma] \, d\xi - [\bar{\epsilon}_E^{1/2} \bar{R}\gamma]_0^\pi - [\bar{M}\gamma']_0^\pi \right) \right\} \qquad (111)$$

By setting the first variation equal to zero, we obtain

$$\int_0^\pi p\zeta' \, d\xi - [\bar{p}\zeta]_0^\pi = 0$$

$$\qquad (112)$$

$$\int_0^\pi [p\eta'\gamma' + (\eta'' - \eta_0'')\gamma'' + \beta(\eta - \eta_0)\gamma - q\gamma - F\delta(\xi - \bar{\xi})\gamma] \, d\xi$$

$$- [\bar{\epsilon}_E^{1/2} \bar{R}\gamma]_0^\pi - [\bar{M}\gamma']_0^\pi = 0$$

Integration by parts yields the following form for Eqs. (112)

$$-\int_0^\pi p'\zeta \, d\xi + [(p - \bar{p})\zeta]_0^\pi = 0$$

$$\int_0^\pi [-(p\eta')' + (\eta'' - \eta_0'')'' + \beta(\eta - \eta_0) - q - F\delta(\xi - \bar{\xi})] \gamma \, d\xi \qquad (113)$$

$$+ \{[p\eta' - (\eta'' - \eta_0'')' - \bar{\epsilon}_E^{1/2}\bar{R}]\gamma\} |_0^\pi + \{[(\eta'' - \eta_0'') - \bar{M}]\gamma'\} |_0^\pi = 0$$

Through the fundamental lemma of the calculus of variations, we obtain from Eqs. (113) equilibrium equations and boundary conditions.

Equilibrium Equations.

$$p' = 0 \qquad (114)$$

$$(\eta'' - \eta_0'')'' - (p\eta')' + \beta(\eta - \eta_0) - q - F\delta(\xi - \bar{\xi}) = 0 \qquad (115)$$

Boundary Conditions.

Either (kinematic)	*Or (Natural)*	
$\epsilon_1\zeta = \delta v = 0$	$p = \bar{p}$	
$\epsilon_2\gamma' = (\delta\eta)' = 0$	$\eta'' - \eta_0'' = \bar{M}$	(116)
$\epsilon_2\gamma = \delta\eta = 0$	$-(\eta'' - \eta_0'')' + p\eta' = \bar{\epsilon}_E^{1/2}\bar{R}$	

Note that, if the supports are immovable, then $\delta v = 0$ or $v(0) = v(\pi) = 0$. Furthermore, if the immovable supports are pinned, $\delta\eta = 0$ or $\eta(0) = \eta(\pi)$

$= 0$, $\eta''(0) = \eta_0''(0)$ and $\eta''(\pi) = \eta_0''(\pi)$. Finally, if the immovable supports are clamped $\eta(0) = \eta(\pi) = 0$, and $\eta'(0) = \eta_0'(0)$ and $\eta'(\pi) = \eta_0'(\pi)$.

When the supports are immovable, the expression for p, Eqs. 105, after an integration over the length, becomes

$$\int_0^\pi p \, d\xi = \frac{1}{2} \int_0^\pi \left[\frac{2v'}{\bar{\epsilon}_E^{1/2}} + (\eta')^2 - (\eta_0')^2 \right] d\xi$$

and

$$p = \frac{1}{2\pi} \int_0^\pi [(\eta')^2 - (\eta_0')^2] \, d\xi \tag{117}$$

Note that this expression uses the fact that $p(\xi) =$ constant, according to the first of the equilibrium equations, Eqs. (114). With this, we may express the total potential, Eq. (108), solely in terms of η and its space-dependent derivatives. The boundary terms vanish for supported ends (either pinned or clamped).

$$U_T = \frac{1}{\pi} \int_0^\pi \left[\frac{1}{2\pi^2} \left\{ \int_0^\pi [(\eta')^2 - (\eta_0')^2] \, d\xi \right\}^2 + 2(\eta'' - \eta_0'')^2 \right.$$
$$\left. + 2\beta(\eta - \eta_0)^2 - 4q(\eta - \eta_0) - F\delta(\xi - \bar{\xi})(\eta - \eta_0) \right] d\xi \tag{118}$$

7.5 THE SINUSOIDAL PINNED ARCH

The problem to be considered here is a low half-sine pinned arch loaded quasistatically by a half-sine spatially distributed load. The initial shape is given by

$$\eta_0 = e \sin \xi \qquad 0 \leq \xi \leq \pi \tag{119}$$

where e is the initial rise parameter. Since $(w_0)_{max} = pe$, then $e = (w_0)_{max}/p$, and if the cross section is rectangular of width l and thickness h, then $p = h/2\sqrt{3}$ and $e = 2\sqrt{3}(w_0)_{max}/h$, which clearly shows that e is a measure of the ratio of the initial maximum rise to the thickness of the arch. The expression for the loading is given by

$$q(\xi) = q_1 \sin \xi \tag{120}$$

The deflection may be represented by an infinite sine series, each term of which satisfies the boundary conditions

$$\eta(\xi) = \eta_0(\xi) + \sum_{n=1}^\infty a_n \sin n\xi \tag{121}$$

Boundary conditions:

$$\eta(0) = \eta(\pi) = 0, \qquad \eta''(0) = \eta''(\pi) = 0 \qquad (122)$$

Substitution of Eq. (121) into the expression for the total potential, Eq. (118), yields

$$U_T = \tfrac{1}{8}\left[\sum_{n=1}^{\infty} n^2 a_n^2 + 2ea_1\right]^2 + \sum_{n=1}^{\infty} n^4 a_n^2 - 2q_1 a_1 + \beta \sum_{n=1}^{\infty} a_n^2 \qquad (123)$$

We are interested in finding, for the entire range of the free parameters β and e, the load at which instability (snapthrough or bifurcation buckling) is possible. This load is called critical load. We find it by first writing the equilibrium equations and then studying the character of these static equilibrium positions (stability in the small).

To find the static equilibrium positions, we use the principle of the stationary value of the total potential, or

$$\frac{\partial U_T}{\partial a_k} = 0 \qquad k = 1, 2, 3, \ldots \qquad (124)$$

This leads to

$$\tfrac{1}{4}\left[\sum_{n=1}^{\infty} n^2 a_n^2 + 2ea_1\right](a_1 + e) + a_1 + \beta a_1 = q_1$$

$$\tfrac{1}{4}\left[\sum_{n=1}^{\infty} n^2 a_n^2 + 2ea_1\right]k^2 a_k + k^4 a_k + \beta a_k = 0, \qquad k = 2, 3, 4, \ldots \qquad (125)$$

There are three possible cases that result from Eqs. (125)

Case I: $a_1 \neq 0$ and $a_k \equiv 0$ for $k = 2, 3, 4, \ldots$.

Case II: $a_1 \neq 0$, $a_m \neq 0$, and $a_k \equiv 0$ for $k = 2, 3, 4, \ldots$, except $m = k$.

Case III: When $\beta = m^2 j^2$, then it is possible that $a_1 \neq 0$, $a_m \neq 0$, $a_j \neq 0$, and $a_k \equiv 0$ for $k = 2, 3, 4, \ldots$, except $k = j$ and m.

Case III will be treated separately. For the first two cases, a more convenient form of the equilibrium equations may be obtained if we introduce the following new parameters:

$$r_1 = a_1 + e, \qquad Q = q_1 + (1 + \beta)e \qquad (126)$$

With these new parameters, the equilibrium equations are

$$\frac{1}{4}[r_1^2 - e^2 + k^2 a_k^2 + 4(1 + \beta)]r_1 = Q$$

$$\left[\frac{k^2}{4}(r_1^2 - e^2 + k^2 a_k^2) + (k^4 + \beta)\right]a_k = 0 \qquad (127)$$

We see from the equilibrium equations, Eqs. (127), that there are two possibilities: (1) $r_1 \neq 0$ and $a_k \equiv 0$, and (2) $r_1 \neq 0$ and $a_k \neq 0$. All the possible positions of static equilibrium are shown in Fig. 7-4. Note that the starting (undeformed) position is A (see Fig. 7-5) and the possibility of the existence of the a_k-mode is present for $e^2 > (4/k^2)(k^4 + \beta)$.

1. If $a_k \equiv 0$ then, from Eqs. (127), the equilibrium equation which also represents the load-deflection curve is

$$r_1^3 - [e^2 - 4(1 + \beta)]r_1 = 4Q \tag{128}$$

We see from Eq. (128) that, for $e \leq 2\sqrt{1 + \beta}$, there is a one-to-one Q-to-r_1 dependence, and there is no possibility of a snapthrough phenomenon. For

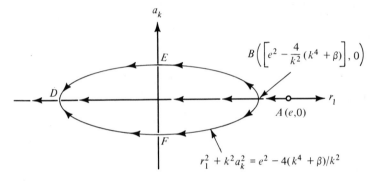

Figure 7-4. Positions of static equilibrium in the (r_1, a_k)-space.

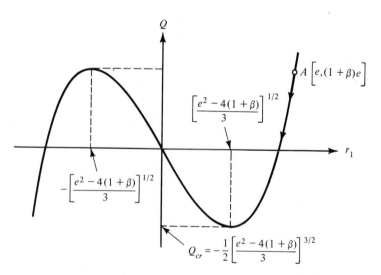

Figure 7-5. Load-deflection graph for symmetric buckling $[e > 2(1 + \beta)\, 1/2]$.

$e < 2\sqrt{1 + \beta}$, since there is a range of Q values for which there are three equilibrium positions for the same Q value, the possibility for a snapthrough phenomenon exists (see Fig. 7-5). It will be shown that for this case the near and far static equilibrium positions are stable and the intermediate one is unstable. When the near static equilibrium position becomes unstable in the small, snapping occurs and the corresponding load is a critical one.

2. If $e^2 > 4(k^4 + \beta)/k^2$, then it is possible for $a_k \neq 0$, and the load-displacement relation is given by the following set of equations in addition to Eq. (128):

$$r_1 \frac{k^2 - 1}{k^2} (\beta - k^2) = Q \tag{129}$$

$$a_k^2 = \frac{1}{k^2}\left[e^2 - \frac{4}{k^2}(k^4 + \beta) - \frac{Q^2 k^4}{(k - 1)^2(\beta - k^2)^2} \right] \tag{130}$$

Thus we see from Eqs. (128), (129), and (130) that the load-displacement relation for the entire range of initial rise parameter values and all possible cases of its relation to the values of k and the modulus of the foundation may be represented by the six graphs of Fig. 7-6.

Critical load shall be defined as the smallest load for which the near static equilibrium position becomes unstable (in the small).

The necessary and sufficient condition for stability (in the small) of the static equilibrium positions given by the roots of Eqs. (128), (129), and (130) is that

$$\frac{\partial^2 U_T}{\partial r_1^2} > 0, \qquad \frac{\partial^2 U_T}{\partial r_1^2} \cdot \frac{\partial^2 U_T}{\partial a_k^2} > \left(\frac{\partial^2 U_T}{\partial r_1 \partial a_k} \right)^2 \tag{131}$$

The expression for U_T obtained by substitution of expressions (126) into Eq. (123) is given by

$$U_T = \tfrac{1}{8}(r_1^2 - e^2 + k^2 a_k^2)^2 - (1 + \beta)(e^2 - r_1^2) + 2Q(e - r_1)$$
$$+ (k^4 + \beta)a_k^2 \tag{132}$$

From Eq. (132), we obtain the following expressions for the second derivatives

$$\frac{\partial^2 U_T}{\partial r_1^2} = \frac{1}{2}[3r_1^2 - e^2 + k^2 a_k^2 + 4(1 + \beta)], \qquad \frac{\partial^2 U_T}{\partial r_1 \partial a_k} = k^2 r_1 a_k$$

$$\frac{\partial^2 U_T}{\partial a_k^2} = \frac{1}{2}[r_1^2 - e^2 + 3k^2 a_k^2]k^2 + 2(k^4 + \beta) \tag{133}$$

First, we investigate the stability of the equilibrium positions that are characterized by the ellipse (see Fig. 7-4). These positions are shown in Figs. 7-6b (AA'), 7-6e (DD'), and 7-6f (EE'). Making use of the equilibrium equa-

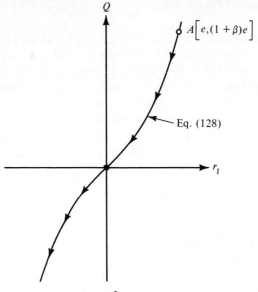

(a) $e \leqslant 2 (1 + \beta)^{1/2}$, $e \leqslant \dfrac{2}{k} (k^4 + \beta)^{1/2}$; $k = 2, 3, \ldots$.

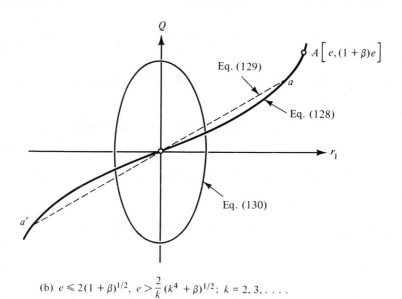

(b) $e \leqslant 2(1 + \beta)^{1/2}$, $e > \dfrac{2}{k} (k^4 + \beta)^{1/2}$; $k = 2, 3, \ldots$.

Figure 7-6. Typical load-deflection (equilibrium) curves.

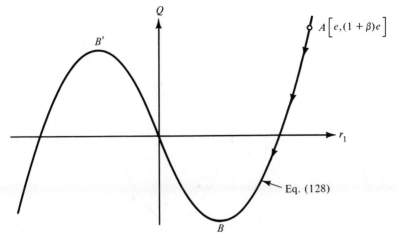

(c) $e > 2(1 + \beta)^{1/2}$, $e \leqslant \dfrac{2}{k}(k^4 + \beta)^{1/2}$; $k = 2, 3, \ldots$.

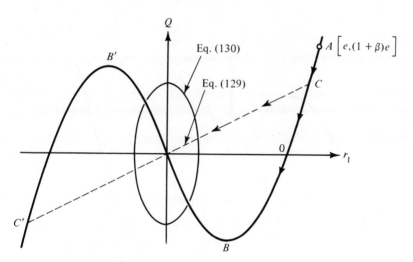

(d) $e > 2(1 + \beta)^{1/2} > \dfrac{2}{k}(k^4 + \beta)^{1/2}$; $k = 2, 3, \ldots$. (this implies $\beta > k^2$)

Figure 7-6. Continued.

tions, the necessary and sufficient conditions for stability, inequalities (131), become

$$r_1^2 + 2\frac{k^2 - 1}{k^2}(\beta - k^2) > 0, \qquad k^4 a_k^2\left[2\frac{k^2 - 1}{k^2}(\beta - k^2)\right] > 0 \quad (134)$$

We see from these inequalities that, if $\beta > k^2$, the ellipse equilibrium posi-

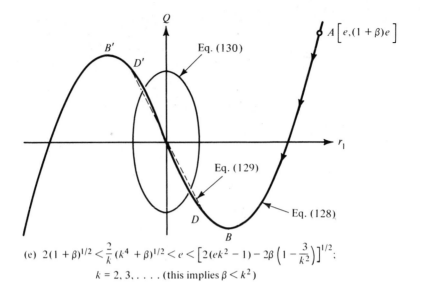

(e) $2(1 + \beta)^{1/2} < \frac{2}{k}(k^4 + \beta)^{1/2} < e < \left[2(ek^2 - 1) - 2\beta\left(1 - \frac{3}{k^2}\right)\right]^{1/2}$;

$k = 2, 3, \ldots$ (this implies $\beta < k^2$)

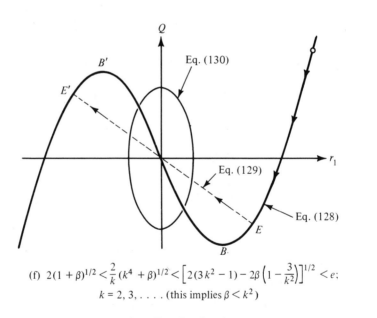

(f) $2(1 + \beta)^{1/2} < \frac{2}{k}(k^4 + \beta)^{1/2} < \left[2(3k^2 - 1) - 2\beta\left(1 - \frac{3}{k^2}\right)\right]^{1/2} < e$;

$k = 2, 3, \ldots$ (this implies $\beta < k^2$)

Figure 7-6. Continued.

tions are stable positions. The test fails only for the $a_k = 0$ positions of the ellipse, but it can easily be shown from the consideration of the third and fourth variations that these two positions are stable.

For the case of $\beta > k^2$, we see that at point A of Fig. 7-6b and point C of Fig. 7-6d there is a possibility of stable bifurcation, [classical buckling—paths AA', Fig. 7-6b, and CC', Fig. 7-6d]. When this happens, the primary state equilibrium position becomes unstable past the bifurcation point [COB of Fig. 7-6d] because the second of Eqs. (131) is not satisfied.

Case III: The existence of a three-mode equilibrium shape.

A three-mode equilibrium position is possible only for $\beta = n^2k^2$ where n and k are differing integers greater than 2. For these distinct values of β, the equilibrium equations, Eq. (125), become

$$[(r_1^2 - e^2 + k^2a_k^2 + n^2a_n^2) + 4(1 + n^2k^2)]r_1 = 4Q$$
$$[k^2(r_1^2 - e^2 + k^2a_k^2 + n^2a_n^2) + 4(k^4 + n^2k^2)]a_k = 0 \qquad (135)$$
$$[n^2(r_1^2 - e^2 + k^2a_k^2 + n^2a_n^2) + 4(n^4 + n^2k^2)]a_n = 0$$

All the possible static equilibrium positions are plotted in Fig. 7-7. Note that the starting point is characterized by $r_1 = e$, and the possibility of the existence of all three modes is realized when $r_1 = \sqrt{e^2 - 4(n^2 + k^2)}$ with the additional condition that $e > 2\sqrt{k^2 + n^2}$.

A typical load-deflection $(Q - r_1)$ plot is shown in Fig. 7-8. The initial (unloaded) position is denoted by point A. As the system is loaded, position B (bifurcation point) is reached, and the system may follow either path BC

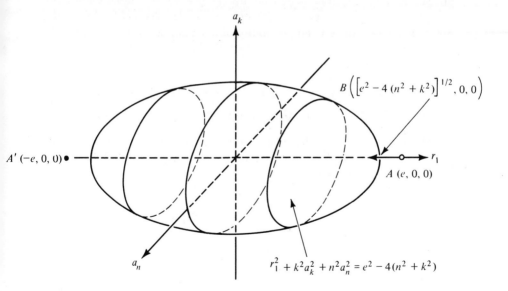

Figure 7-7. Positions of static equilibrium in the (r_1, a_k, a_n)-space.

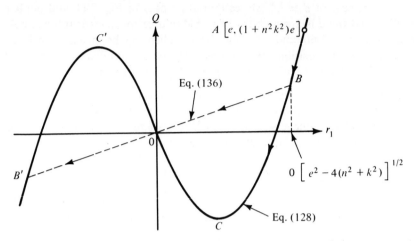

Figure 7-8. Typical load-deflection curves for the three-mode case.

[Eq. (128)] or path BOB' which is characterized by

$$Q = [(1 + n^2k^2) - (n^2 + k^2)]r_1 \tag{136}$$

Note that, since $n^2k^2 > n^2 + k^2$, the slope of the BOB' curve is positive as shown in Fig. 7-8.

It is shown next that the equilibrium positions on the ellipsoid (Fig. 7-7) are stable, and therefore points on BB' (Fig. 7-8) are stable equilibrium positions. Finally, since path BC becomes unstable, classical buckling takes place. Therefore, there is no possibility of a snapping phenomenon for this case.

The total potential for this case is

$$U_T = \tfrac{1}{8}(r_1^2 - e^2 + k^2a_k^2 + n^2a_n^2)^2 - (1 + n^2k^2)(e^2 - r_1^2) \\ + 2Q(e - r_1) + (n^2 + k^2)(k^2a_k^2 + n^2a_n^2) \tag{137}$$

The necessary and sufficient condition for stability of the equilibrium positions on the ellipsoid is that the following determinant and all its principal minors (dashed lines) be positive definite

$$\begin{vmatrix} \dfrac{\partial^2 U_T}{\partial r_1^2} & \dfrac{\partial^2 U_T}{\partial r_1 \partial a_k} & \dfrac{\partial^2 U_T}{\partial r_1 \partial a_n} \\[2mm] \dfrac{\partial^2 U_T}{\partial r_1 \partial a_k} & \dfrac{\partial^2 U_T}{\partial a_k^2} & \dfrac{\partial^2 U_T}{\partial a_k \partial a_n} \\[2mm] \dfrac{\partial^2 U_T}{\partial r_1 \partial a_n} & \dfrac{\partial^2 U_T}{\partial a_k \partial a_n} & \dfrac{\partial^2 U_T}{\partial a_n^2} \end{vmatrix} > 0 \tag{138}$$

Use of the expression for the total potential and the equilibrium equations leads to the fact that the principal minors are positive definite, but the determinant is identically equal to zero; therefore the test fails.

Checking the higher variations, we can show that $\delta^3 U_T \equiv 0$ and $\delta^4 U_T$ (fourth variation) is positive definite because all of the fourth-order derivatives are zero except

$$\frac{\partial^4 U_T}{\partial r_1^4} = 3, \qquad \frac{\partial^4 U_T}{\partial r_1^2 \partial a_k^2} = k^2, \qquad \frac{\partial^4 U_T}{\partial r_1^2 \partial a_n^2} = n^2,$$

$$\frac{\partial^4 U_T}{\partial a_k^4} = 3k^4, \qquad \frac{\partial^4 U_T}{\partial a_n^4} = 3n^2 \tag{139}$$

Because of this, the equilibrium positions on the ellipsoid are stable, and path BB' (Fig. 7-8) is a stable path. Finally, the primary path, BC, becomes unstable and the model exhibits classical buckling (adjacent equilibrium) at point B.

7.5-1 Critical Loads

Note from Eqs. (134) that if $\beta \geq k^2$, $k = 2, 3, \ldots$, there is no possibility of snapping but there is bifurcation buckling. Therefore we must consider the following two ranges for β values separately.

Range 1, $\beta < 4$. Snapping is possible and the following cases must be distinguished:

1. If $2\sqrt{1 + \beta} < e < \sqrt{16 + \beta}$ [see Fig. 7-6c], then the system will reach point B and snap through an a_1-only mode. The critical load for this case is obtained from Eq. (128) with $r_1 = [e^2 - 4(1 + \beta)]^{1/2}/3^{1/2}$

$$Q_{cr} = -\frac{1}{2}\left[\frac{e^2 - 4(1 + \beta)}{3}\right]^{3/2};$$

$$q_{1_{cr}} = -(1 + \beta)e - \frac{1}{2}\left[\frac{e^2 - 4(1 + \beta)}{3}\right]^{3/2} \tag{140}$$

2. If $2\sqrt{1 + \beta} < \sqrt{16 + \beta} < e < \sqrt{22 - \beta/2}$, then, although an unstable bifurcation through mode a_2 is possible, the system will snap initially through an a_1-mode, because during the loading process point B will be reached before point D [see Fig. 7-6e]. In this case, the critical load is still given by Eq. (140).

3. If $e > \sqrt{22 - \beta/2} > \sqrt{16 + \beta} > 2\sqrt{1 + \beta}$, then snapping will take place through an a_2-mode [point E of Fig. 7-6f], and the critical load is

given by Eq. (129) with r_1 equal to the value corresponding at the bifurcation point or

$$q_{1_{cr}} = -(1 + \beta)e - 3\left(1 - \frac{\beta}{4}\right)[e^2 - (16 + \beta)]^{1/2} \qquad (141)$$

Range 2, $\beta \geq 4$. Stable bifurcational buckling takes place and the load at the bifurcation point is given by (the subscript cl.B. means classical buckling)

$$Q_{cl. B.} = \frac{k^2 - 1}{k^2}(\beta - k^2)\left[e^2 - \frac{4}{k^2}(k^4 + \beta)\right]^{1/2} \qquad (142)$$

We see from this expression that the smallest bifurcation load and the corresponding mode of deformation depend on the value of the modulus of foundation, β.

For $4 \leq \beta \leq 36$, $k = 2$ and $Q_{cl. B.} = \frac{3}{4}(\beta - 4)[e^2 - (18 + \beta)]^{1/2}$

For $36 \leq \beta \leq 144$, $k = 3$ and $Q_{cl. B.} = \frac{8}{9}(\beta - 9)[e^2 - \frac{4}{9}(81 + \beta)]^{1/2}$

For $144 \leq \beta \leq 400$, $k = 4$ and $Q_{cl. B.} = \frac{15}{16}(\beta - 16)[e^2 - \frac{1}{4}(256 + \beta)]^{1/2}$

Note that at $\beta = k^2(k + 1)^2$, bifurcation occurs either through an a_k-mode or through an a_{k+1}-mode or a combination of a_k- and a_{k+1}-modes (the three-mode case).

Numerical results are presented graphically in Figs. 7-9 and 7-10. For $\beta = 0$, the results reduce to those reported in Refs. 12 and 13. For $\beta = 2$, if $e \leq \sqrt{12}$, there is no possibility of snapthrough. If $\sqrt{12} < e < \sqrt{21}$, the critical load is given by

$$q_{1_{cr}} = -3e - \frac{1}{2}\left(\frac{e^2 - 12}{3}\right)^{3/2}$$

and snapping occurs through a limit point instability (top-of-the-knee). If $e < \sqrt{21}$ the critical load is given by

$$q_{1_{cr}} = -3e - \frac{9}{4}(e^2 - 18)^{1/2}$$

and snapping takes place through an unstable bifurcation. These results are shown in Fig. 7-9 as $(-q_{cr})$ versus the initial rise parameter.

For $\beta > 4$ the results are presented in Fig. 7-10 as classical buckling load versus the initial rise parameter. Note that when $\beta = 36$, the stable branch is characterized by the a_2-mode alone or a_3-mode alone, or a combined a_2, a_3-mode. This phenomenon is similar to the pinned straight column on an elastic foundation. When $\beta = k^2(k^2 + 1)$, the column can buckle in either k or $k + 1$ half-sine waves (see Fig. 6-3).

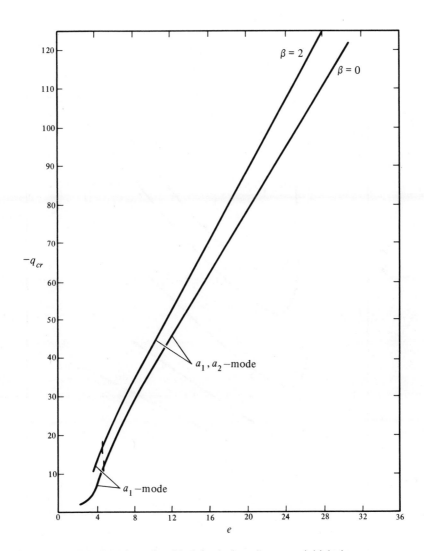

Figure 7-9. Snapthrough critical load, $(-q_{cr})$, versus initial rise parameter, $e(\beta < 4)$.

7.6   THE LOW ARCH BY THE TREFFTZ CRITERION

According to the Trefftz criterion, we must set the first variation of the second variation of the total potential equal to zero at the critical condition. In order to have a convenient expression for the second variation, we shall use one of

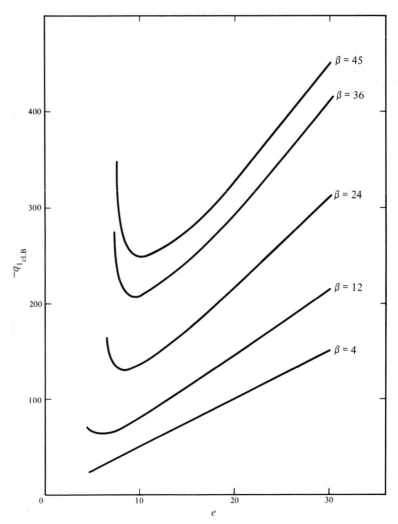

Figure 7-10. Classical buckling load, $(-q_{cl.B})$, versus initial rise parameter, $e(\beta \geq 4)$.

the conditions for equilibrium or $p = $ constant, Eq. (114). Through Eqs. (105), (114), and (117)

$$\frac{2v'}{\bar{\epsilon}_E^{1/2}} + (\eta')^2 - (\eta_0')^2 = \frac{1}{\pi} \int_0^\pi [(\eta')^2 - (\eta_0')^2]\, d\xi \qquad (143)$$

If we take the variations of both sides, we have

$$\pi\left[\frac{2v'}{\bar{\epsilon}_E^{1/2}} + 2\epsilon_1\frac{\zeta'}{\bar{\epsilon}_E^{1/2}} + (\eta')^2 - (\eta_0')^2 + 2\epsilon_2\eta'\gamma' + \epsilon_2^2(\gamma')^2\right]$$
$$= \int_0^\pi [(\eta')^2 - (\eta_0')^2]\, d\xi + \int_0^\pi [2\epsilon_2\eta'\gamma' + \epsilon_2^2(\gamma')^2]\, d\xi \qquad (144)$$

Now, making use of Eq. (143), we obtain

$$\pi\left[2\epsilon_1\frac{\zeta'}{\bar{\epsilon}_E^{1/2}} + 2\epsilon_2\eta'\gamma' + \epsilon_2^2(\gamma')^2\right] = \int_0^\pi [2\epsilon_2\eta'\gamma' + \epsilon_2^2(\gamma')^2]\, d\xi \qquad (145)$$

Squaring both sides and integrating over π does not violate the validity of Eq. (145). Thus

$$\int_0^\pi \left[\epsilon_1\frac{\zeta'}{\bar{\epsilon}_E^{1/2}} + \epsilon_2\eta'\gamma' + \frac{1}{2}\epsilon_2^2(\gamma')^2\right]^2 d\xi = \frac{1}{4\pi}\left[\int_0^\pi \{2\epsilon_2\eta'\gamma' + \epsilon_2^2(\gamma')^2\}\, d\xi\right]^2$$
$$(146)$$

Next we return to Eq. (110) which can be written as

$$\Delta U_T = \delta^1 U_T + \frac{1}{\pi}\int_0^\pi \left\{2\left[\frac{\epsilon_1^2}{\bar{\epsilon}_E}(\zeta')^2 + 2\frac{\epsilon_1\epsilon_2}{\bar{\epsilon}_E^{1/2}}\eta'\zeta'\gamma' + \epsilon_2^2(\eta')^2(\gamma')^2\right.\right.$$
$$\left.+ \frac{\epsilon_1\epsilon_2^2}{\bar{\epsilon}_E^{1/2}}(\gamma')^2\zeta' + \epsilon_2^3\eta'(\gamma')^3 + \frac{1}{4}\epsilon_2^4(\gamma')^4\right]$$
$$+ 2p\epsilon_2^2(\gamma')^2 + 2\epsilon_2^2(\gamma'')^2 + 2\beta\epsilon_2^2\gamma^2\right\}\, d\xi \qquad (147)$$

Rearranging the terms of the integrand on the right-hand side, we obtain

$$\Delta U_T = \delta^1 U_T + \frac{1}{\pi}\int_0^\pi \left\{2\left[\epsilon_1\frac{\zeta'}{\bar{\epsilon}_E^{1/2}} + \epsilon_2\eta'\gamma' + \frac{1}{2}\epsilon_2(\gamma')^2\right]^2\right.$$
$$\left.+ 2p\epsilon_2^2(\gamma')^2 + 2\epsilon_2^2(\gamma'')^2 + 2\beta\epsilon_2^2\gamma^2\right\}\, d\xi \qquad (148)$$

Use of Eq. (146) for the first term in the integrand results in the following form for Eq. (148):

$$\Delta U_T = \delta^1 U_T + \frac{1}{2\pi^2}\left[\int_0^\pi \{2\epsilon_2\eta'\gamma' + \epsilon_2^2(\gamma')^2\}\, d\xi\right]^2$$
$$+ \frac{2\epsilon_2^2}{\pi}\int_0^\pi [p(\gamma')^2 + (\gamma'')^2 + \beta\gamma^2]\, d\xi \qquad (149)$$

Performing the indicated operations and grouping terms on the right-hand side according to powers of ϵ_2, we have

$$\Delta U_T = \delta^1 U_T + \frac{2\epsilon_2^2}{\pi} \int_0^\pi \left[\frac{1}{\pi} \left(\int_0^\pi \eta'\gamma' \, d\xi \right) \eta'\gamma' + p(\gamma')^2 + (\gamma'')^2 + \beta\gamma^2 \right] d\xi$$

$$+ \frac{2\epsilon_2^3}{\pi} \int_0^\pi \eta'\gamma' \, d\xi \int_0^\pi (\gamma')^2 \, d\xi + \frac{\epsilon_2^4}{2\pi} \left[\int_0^\pi (\gamma')^2 \, d\xi \right]^2 \qquad (150)$$

From Eq. (150), it is clear that

$$\delta^2 U_T = \frac{2\epsilon_2^3}{\pi} \int_0^\pi \left[\frac{1}{\pi} \left(\int_0^\pi \eta'\gamma' \, d\xi \right) \eta'\gamma' + p(\gamma')^2 + (\gamma'')^2 + \beta(\gamma)^2 \right] d\xi \qquad (151)$$

Next, let $(\pi/2\epsilon_2^2)\delta^2 U_T = V[\gamma]$ and find the first variation. Let $\theta(\xi)$ be a kinematically admissible function of ξ (same as γ), and ϵ_3 small constant; then

$$V[\gamma + \epsilon_3\theta] = \int_0^\pi \left[\left(\frac{1}{\pi} \int_0^\pi \eta'(\gamma' + \epsilon_3\theta') \, d\xi \right) \eta'(\gamma' + \epsilon_3\theta') + \right.$$
$$\left. p(\gamma' + \epsilon_3\theta')^2 + (\gamma'' + \epsilon_3\theta'')^2 + \beta(\gamma + \epsilon_3\theta)^2 \right] d\xi \qquad (152)$$

Performing the indicated operations and grouping terms according to powers of ϵ_3, we have

$$V[\gamma + \epsilon_3\theta] = \int_0^\pi \left[\left(\frac{1}{\pi} \int_0^\pi \eta'\gamma' \, d\xi \right) \eta'\gamma' + p(\gamma')^2 + (\gamma'')^2 + \beta\gamma^2 \right] d\xi$$

$$+ \epsilon_3 \int_0^\pi \left[\left(\frac{1}{\pi} \int_0^\pi \eta'\theta' \, d\xi \right) \eta'\gamma' + \left(\frac{1}{\pi} \int_0^\pi \eta'\gamma' \, d\xi \right) \eta'\theta' \right.$$

$$\left. + 2p\gamma'\theta' + 2\gamma''\theta'' + 2\beta\gamma\theta \right] d\xi$$

$$+ \epsilon_3^2 \int_0^\pi \left[\left(\frac{1}{\pi} \int_0^\pi \eta'\theta' \, d\xi \right) \eta'\theta' + p(\theta')^2 + (\theta'')^2 + \beta\theta^2 \right] d\xi$$
$$\qquad (153)$$

It is clear from Eq. (153) that the first variation in V, which must vanish, is given by

$$\delta^1 V = 2\epsilon_3 \int_0^\pi \left[\left(\frac{1}{\pi} \int_0^\pi \eta'\gamma' \, d\xi \right) \eta'\theta' + p\gamma'\theta' + \gamma''\theta'' + \beta\gamma\theta \right] d\xi = 0 \qquad (154)$$

Integration by parts yields

$$\left[\left(-\gamma''' + p\gamma' + \left\{\frac{1}{\pi}\int_0^\pi \eta'\gamma'\,d\xi\right\}\eta'\right)\theta\right]_0^\pi + [\gamma''\theta']_0^\pi$$

$$+ \int_0^\pi \left[\gamma'''' - p\gamma'' - \left(\frac{1}{\pi}\int_0^\pi \eta'\gamma'\,d\xi\right)\eta'' + \beta\gamma\right]\theta\,d\xi = 0 \qquad (155)$$

Since θ and γ are kinematically admissible, the first of the boundary terms is zero, Eq. (155), as long as the arch is supported, $\eta(0) = \eta(\pi) = 0$, and regardless of whether the support is pinned or clamped. Furthermore, the necessary condition for the vanishing of the first variation is the following differential equation in η and γ, and boundary conditions

$$\gamma'''' - p\gamma'' - \left(\frac{1}{\pi}\int_0^\pi \eta'\gamma'\,d\xi\right)\eta'' + \beta\gamma = 0 \qquad (156)$$

$$\begin{array}{cc} \textit{Either} & \textit{Or} \\ \theta' = 0 \quad (\delta\gamma' = 0) & \gamma'' = 0 \end{array} \qquad (156a)$$

Since

$$\int_0^\pi \eta'\gamma'\,d\xi = -\int_0^\pi \eta''\gamma\,d\xi + [\eta'\gamma]_0^\pi$$

Eq. (156) becomes

$$\gamma'''' - p\gamma'' + \left(\frac{1}{\pi}\int_0^\pi \eta''\gamma\,d\xi\right)\eta'' + \beta\gamma = 0 \qquad (157)$$

In summary, we conclude that the response of the arch (primary path), $\eta(\xi)$, the critical load, q_{cr}, and the buckling mode, $\gamma(\xi)$, are established through the simultaneous solution of Eqs. (115) and (157) subject to the proper boundary conditions. This is demonstrated in the next section where we consider the pinned half-sine arch under a half-sine spatial distribution of the load.

7.6-1 The Sinusoidal Arch

Consider a half-sine arch under a half-sine load pinned at both ends. It has been demonstrated in Section 7.5 that Eq. (115) is satisfied (with $\beta = 0$) if

$$\begin{aligned} \tfrac{1}{4}(r_1^2 - e^2 + k^2 a_k^2)r_1 + r_1 &= Q \\ [\tfrac{1}{4}(r_1^2 - e^2 + k^2 a_k^2) + k^2]a_k &= 0 \end{aligned} \qquad (127)$$

where

$$\eta_0 = e \sin \xi, \qquad q = q_1 \sin \xi$$
$$r_1 = a_1 + e, \qquad Q = q_1 + e$$
$$\eta = (e + a_1) \sin \xi + a_k \sin k\xi \qquad (158)$$
$$p = \frac{1}{2\pi} \int_0^\pi [(\eta')^2 - (\eta_0')^2] \, d\xi$$

Substitution of the needed expressions in Eqs. (158) into Eq. (157) yields

$$\gamma'''' - \frac{1}{4}(r_1^2 - e^2 + k^2 a_k^2)\gamma'' + (r_1 \sin \xi + k^2 a_k \sin k\xi) \cdot$$
$$\cdot \frac{1}{\pi} \int_0^\pi (r_1 \sin \xi + k^2 a_k \sin k\xi)\gamma \, d\xi = 0 \qquad (159)$$

If we let $\gamma = \sum_{m=1}^{\infty} A_m \sin m\xi$, we note that every term in the series for γ is kinematically admissible and satisfies the boundary conditions, $\gamma''(0) = \gamma''(\pi) = 0$. Substitution into Eq. (157), because of the linear independence of the functions $\sin m\xi$, yields

$$m = 1: \qquad A_1[1 + \tfrac{1}{4}(r_1^2 - e^2 + k^2 a_k^2)] + \tfrac{1}{2}r_1(r_1 A_1 + k^2 a_k A_k) = 0 \qquad (160)$$

$$m = k:$$

$$k^2 A_k \left[k^2 + \frac{1}{4}(r_1^2 - e^2 + k^2 a_k^2) \right] + \frac{k^2}{2} a_k(r_1 A_1 + k^2 a_k A_k) = 0 \qquad (161)$$

$$m \neq 1, k: \qquad m^2 A_m[m^2 + \tfrac{1}{4}(r_1^2 - e^2 + k^2 a_k^2)] = 0 \qquad (162)$$

From Eq. (162) it is clear that, if one A_m is not zero, all other A_m must be zero. Furthermore, it can be shown that all A_m must be zero for a meaningful solution to exist (see Problems at the end of the chapter).

Now we can proceed to find the position of the bifurcation point (r_{1cr}, a_{kcr}, Q_{cr}) and the buckling mode (A_1, A_k). This is done by seeking the simultaneous satisfaction of Eqs. (127), (160), and (161).

$$\tfrac{1}{4}(r_{1cr}^2 - e^2 + k^2 a_{kcr}^2)r_{1cr} + r_{1cr} = Q_{cr}$$
$$[\tfrac{1}{4}(r_{1cr}^2 - e^2 + k^2 a_{kcr}^2) + k^2]a_{kcr} = 0$$
$$A_1[1 + \tfrac{1}{4}(r_{1cr}^2 - e^2 + k^2 a_{kcr}^2)] + \tfrac{1}{2}a_{kcr}(r_{1cr}A_1 + k^2 a_{kcr}A_k) = 0 \qquad (163)$$
$$A_k[k^2 + \tfrac{1}{4}(r_{1cr}^2 - e^2 + k^2 a_{kcr}^2)] + \tfrac{1}{2}a_{kcr}(r_{1cr}A_1 + k^2 a_{kcr}A_k) = 0$$

Equations (163) denote a system of four equations in five unknowns; r_{1cr} and a_{kcr} are the positions of the bifurcation point on the r_1-a_k equilibrium positions space (see Fig. 7-4); Q_{cr} is the corresponding critical load (limit point or bifurcation); and A_1, A_k are the amplitudes of the buckling mode (both cannot be determined uniquely).

Recognizing that, as the load is increased from zero, the primary path is associated with an r_1-only mode ($a_k \equiv 0$) and that through the Trefftz criterion we may obtain both the limit point as well as the bifurcation point, we must consider only the case of $r_1 \neq 0$ and $a_k \equiv 0$. There is no need to consider the case of $r_1 \neq 0$ and $a_k \neq 0$ which may also satisfy the equilibrium equations. (See cases I and II of Section 7.5.) With this, Eqs. (163) become

$$\tfrac{1}{4}(r_{1_{cr}}^2 - e^2)r_{1_{cr}} + r_{1_{cr}} = Q_{cr} \tag{164}$$

$$A_1[1 + \tfrac{1}{4}(r_{1_{cr}}^2 - e^2) + \tfrac{1}{2}r_{1_{cr}}^2] = 0 \tag{165}$$

$$A_k[k^2 + \tfrac{1}{4}(r_{1_{cr}}^2 - e^2)] = 0 \tag{166}$$

Equations (164) through (166) suggest two possible solutions: (1) $A_1 \neq 0$, $A_k = 0$ and (2) $A_1 = 0$, $A_k \neq 0$. In the case for which $A_1 \neq 0$, and $A_k \equiv 0$, from Eq. (165) we have

$$r_{1_{cr}} = \left(\frac{e^2 - 4}{3}\right)^{1/2} \tag{167}$$

Substitution of this expression for $r_{1_{cr}}$, Eq. (167), into Eq. (164) gives

$$Q_{cr} = -\frac{1}{2}\left(\frac{e^2 - 4}{3}\right)^{3/2} \tag{168}$$

These are the results obtained in Section 7.5 for top-of-the-knee buckling (limit point stability). It is clear that buckling is possible for arches with $e > 2$.

In the case for which $A_1 = 0$ and $A_k \neq 0$, from Eq. (166) we obtain

$$r_{1_{cr}} = (e^2 - 4k^2)^{1/2} \tag{169}$$

Substitution into Eq. (164) yields

$$Q_{cr} = -(k^2 - 1)(e^2 - 4k^2) \tag{170}$$

Minimization of Q_{cr} with respect to integer values of k shown that $k = 2$ and

$$r_{1_{cr}} = (e^2 - 16)^{1/2}$$
$$Q_{cr} = -3(e^2 - 16)^{1/2} \tag{171}$$

These are identical to the results obtained in the case of bifurcation snapping in Section 7.5. Note that antisymmetric buckling is possible if $e > 4$. Note also that if the limit point is reached before the bifurcation point (see Fig. 7-6), Q_{cr} is still given by Eq. (168):

$$\left(\frac{e^2 - 4}{3}\right)^{1/2} > (e^2 - 16)^{1/2}$$

$$e < \sqrt{22}$$

Thus, for $2 < e < \sqrt{22}$

$$Q_{cr} = -\frac{1}{2}\left(\frac{e^2 - 4}{3}\right)^{3/2}$$

and for $e > \sqrt{22}$

$$Q_{cr} = -3(e^2 - 16)^{1/2}$$

PROBLEMS

1. Find the critical load for a thin ring under uniform pressure (load case I) when one section of the ring is fixed in space (say, at $\theta = 0$, $v = 0$, $\varphi = 0$, and $w = 0$).

2. Consider an arch as shown in Fig. 7-2 and labeled "original" and find p_{cr} (load case I). As boundary conditions, assume that the shear in the radial direction is zero instead of the displacement.

3. Consider a clamped arch on rollers (similar to Fig. 7-2 labeled "original") and find p_{cr} (load case I).

4. Using the alternate solution (Section 7.4.2), find p_{cr} for a pinned arch for load case II.

5. Using the alternate solution (Section 7.4.2) find p_{cr} for a pinned arch for load case III.

6. Show that there is no meaningful solution if $A_m \neq 0$ in Eq. (162).

7. Using the Trefftz criterion approach (Section 7.6.1), analyze a pinned half-sine arch under a half-sine loading and resting on an elastic foundation.

REFERENCES

1. TIMOSHENKO, S. P., and GERE, J., *Theory of Elastic Stability*, McGraw-Hill Book Co., New York, 1961, pp. 287–293.

2. BORESI, A., "A Refinement of the Theory of Buckling of Rings under Uniform Pressure," *J. Appl. Mech.*, Vol. 22, pp. 95–102, 1955.

3. WASSERMAN, E., "The Effect of the Behavior of the Load on the Frequency of Free Vibrations of a Ring," NASA TT-F-52, 1961.

4. WEMPNER, G., and KESTI, N., "On the Buckling of Circular Arches and Rings," *Proceedings, Fourth U.S. National Congress of Applied Mechanics*, ASME, Vol. 2, pp. 843–852, 1962.

5. SMITH, C. V., JR., and SIMITSES, G. J., "Effect of Shear and Load Behavior on Ring Stability," *Proc. ASCE*, EM3, pp. 559–569, 1969.

6. SANDERS, J. L., JR., "Nonlinear Theories for Thin Shells," *Quart. Appl. Math,* Vol. 21, pp. 21–36, 1963.

7. SINGER, J., and BABCOCK, C. O., "On the Buckling of Rings under Constant Directional and Centrally Directed Pressure," *J. Appl. Mech.*, Vol. 37, No. 1, pp. 215–218, 1970.

8. MARGUERRE, K., "Die Durchschlagskraft eines Schwach Gekrummten Balken," *Sitz. Berlin Math. Gess.*, Vol. 37, p. 92, 1938.

9. TIMOSHENKO, S. P., "Buckling of Curved Bars with Small Curvature," *J. Appl. Mech.*, Vol. 2, No. 1, p. 17, 1935.

10. BIEZENO, C. B., "Das Durchschlagen eines Schwach Gekrummten Stabes," *Zeitschrift Ange. Math. und Mekh.*, Vol. 18, p. 21, 1938.

11. NAVIER, "Résumé des Leçons sur L'Application de la Méchanique," 2nd ed., p. 273, Paris, 1833.

12. FUNG, Y. C., and KAPLAN, A., "Buckling of Low Arches or Curved Beams of Small Curvature," NACA TN 2840, 1952.

13. HOFF, N. J., and BRUCE, V. G., "Dynamic Analysis of the Buckling of Laterally Loaded Flat Arches," *J. Math and Phys.*, Vol. XXXII, No. 4, 1954.

14. GJELSVIK, A., and BODNER, S. R., "The Energy Criterion and Snap Buckling of Arches," *J. Eng. Mech. Div.*, ASCE, Vol. 88, EM5, p. 87, 1962.

15. SCHREYER, H. L., and MASUR, E. F., "Buckling of Shallow Arches," *J. Eng. Mech. Div.*, ASCE, Vol. 92, EM4, p. 1, 1966.

16. MASUR, E. F., and LO, D. L. C., "The Shallow Arch-General Buckling, Post-buckling, and Imperfection Analysis," *J. Struct. Mech.*, Vol. 1, No. 1, p. 91, 1972.

17. FRANCIOSI, V., AUGUSTI, G., and SPARACIO, R., "Collapse of Arches Under Repeated Loading," *J. Struct. Div.*, ASCE, Vol. 90, STI, p. 165, 1964.

18. ONAT, E. T., and SHU, L. S., "Finite Deformation of a Rigid Perfectly Plastic Arch," *J. Appl. Mech.*, Vol. 29, No. 3, p. 549, 1962.

19. LEE, H. N., and MURPHY, L. M., "Inelastic Buckling of Shallow Arches," *J. Eng. Mech. Div.*, ASCE, Vol. 94, EM1, p. 225, 1968.

20. ROORDA, J., "Stability of Structures with Small Imperfections," *J. Eng. Mech. Div.*, ASCE, Vol. 91, EM1, p. 87, 1965.

21. SIMITSES, G. J., "Snapping of Low Pinned Arches on an Elastic Foundation," *J. Appl. Mech.*, Vol. 40, No. 3, p. 741, 1973.

8

NONCONSERVATIVE SYSTEMS
AND DYNAMIC BUCKLING

8.1 PRELIMINARY REMARKS

All of the previous chapters have dealt with the stability of conservative elastic structural systems under static loads. If we still constrain our discussion to linearly elastic material behavior, we realize that such systems are acted upon by loads which are always time-dependent. The question then arises: "What is a static load?" We shall define loads to be static if their rate of application is so small as not to induce any appreciable dynamic response in the structure. Realizing that the rate of application may vary in building the load to its final value, a good measure of the above definition is to say that, if the load were to be applied at its maximum rate, the time required to reach its final value must be greater than the period of small free oscillations of the structure. Many authors prefer to call such loads quasistatic rather than static.

If we use this definition for static loads, all other loads shall be called dynamic. At this point it is clear that dynamic loads can be both time-dependent as well as time-independent. An example of the latter case is a load applied suddenly with constant magnitude and extremely large (infinite) duration. Furthermore, a time-independent load may be velocity-dependent or velocity-independent. A classification of loads and reactions, when dealing with all mechanical systems, is given by Ziegler (Ref. 1).

A system is conservative when subjected to conservative forces (see

217

Chapter 1). It is possible in some cases to have nonconservative systems under static loads and conservative systems under dynamic loads. In this Chapter, we will consider some examples of both in order to motivate the student for further study in the area of stability of elastic systems under nonconservative static loads and under dynamic loads both conservative and nonconservative. Excellent sources for such further studies are the works of Leipholz (Ref. 2), Bolotin (Ref. 3), Herrmann (Ref. 4), Bolotin (Ref. 5), and others (Ref. 6).

8.2 NONCONSERVATIVE SYSTEMS UNDER STATIC FORCES

A number of solutions have been reported in the open literature dealing with the problem of buckling of elastic systems under nonconservative forces. A comprehensive review may be found in Ref. 7. Some of the basic problems of nonconservative loads are the *follower-force* problems. In such problems, the forces follow the deformations of the body in some manner and the work done in reaching a final position is path-dependent. Consider, for example, the mechanical system shown in Fig. 8-1. It consists of two rigid bars connected with a rotational spring at B and at A as shown. The force P remains always in the direction of bar BC. Consider the two lengths equal, L, and the angles φ and θ small.

The final position $AB'C'$, Fig. 8-1, may be reached in different ways. Two possible paths are shown in Figs. 8-2 and 8-3. In both loading processes, we assume that we first built the load to its final value P while the two bars remain in the upright position. Since the bars are rigid, the work done is

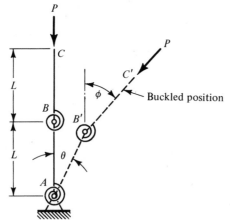

Figure 8-1. An example of a non-conservative force.

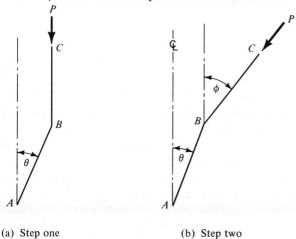

(a) Step one (b) Step two

Figure 8-2. Loading process *A*.

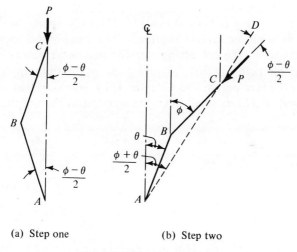

(a) Step one (b) Step two

Figure 8-3. Loading process *B*.

zero. Then in loading process *A*, we rotate bar *AB* through θ while bar *BC* remains upright. The work done is

$$W_A = PL(1 - \cos \theta) = \tfrac{1}{2} PL\theta^2 \tag{1}$$

Then we rotate bar *BC* and force *P* through an angle φ. The additional work done is zero, since the force *P* is normal to the path in this second step. Thus the work in loading process *A* is $\tfrac{1}{2} PL\theta^2$.

In loading process *B*, we first allow point *C* to move along the upright position so that the angle between the upright and *AB* is $(\varphi - \theta)/2$ as shown.

The work done during step 1 is

$$W_B = P2L[1 - \cos(\varphi - \theta)]/2 \left.\vphantom{\begin{matrix}1\\1\end{matrix}}\right\}$$
$$= \tfrac{1}{4}PL(\varphi - \theta)^2 \qquad\qquad (2)$$

To reach the final position shown as $AB'C'$ in Fig. 8-1, we must rotate ABC in Fig. 8-3 in a clockwise manner by an angle $(\varphi + \theta)/2$. If the force P is kept radial (along ACD), the additional work done is zero. Finally, we rotate the force by an angle $(\varphi - \theta)/2$ in a clockwise direction. Thus, the total work done during loading process B is $PL(\varphi - \theta)^2/4$.

We conclude that the work done by force P is path-dependent. Thus the force and consequently the system is nonconservative.

8.2-1 · The Beck Problem

Consider a cantilever (see Fig. 8-4) under an axial follower-force. This problem is called the Beck problem because Beck (Ref. 8) was the first to achieve a correct solution.

Because the system is nonconservative, the only approach that can be used is the dynamic or kinetic approach. The equilibrium approach may or may not lead to the correct answer for nonconservative systems. We demonstrate this point by first employing the equilibrium approach. If we follow the procedure outlined in Chapter 3, by assuming that the rotation at the free end, $w_{,x}(L)$, is small, then the solution to the buckling equation is given by

$$w(x) = A_1 \sin kx + A_2 \cos kx + A_3 x + A_4$$

The boundary conditions are

$$\left.\begin{aligned} w(0) = w_{,x}(0) &= 0 \\ w_{,xx}(L) &= 0 \end{aligned}\right\} \qquad (3)$$

and

$$-[EIw_{,xxx}(L) + \bar{P}w_{,x}(L)] = -\bar{P}w_{,x}(L) \Longrightarrow w_{,xxx}(L) = 0$$

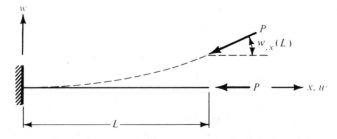

Figure 8-4. The Cantilever under a follower force.

The characteristic equation is the following determinant

$$
\begin{vmatrix}
0 & 1 & 0 & 1 \\
k & 0 & 1 & 0 \\
\sin kL & \cos kL & 0 & 0 \\
\cos kL & -\sin kL & 0 & 0
\end{vmatrix} = 0
\tag{4}
$$

Therefore the only solution is the trivial solution.

If we use the kinetic approach, the equation of motion (see Chapter 3) is

$$
w_{,xxxx} + k^2 w_{,xx} + \frac{m}{EI} w_{,tt} = 0
\tag{5}
$$

Let the solution be of the form

$$
w(x, t) = e^{i\omega t} g(x)
\tag{6}
$$

Substitution into Eq. (5) yields the following ordinary differential equation in $g(x)$:

$$
\frac{d^4 g}{dx^4} + k^2 \frac{d^2 g}{dx^2} - \frac{m\omega^2}{EI} g(x) = 0
\tag{7}
$$

The solution to Eq. (7) is given by

$$
g(x) = B_1 \sin \alpha x + B_2 \cos \alpha x + B_3 \sinh \beta x + B_4 \cosh \beta x
\tag{8}
$$

where

$$
\left.
\begin{aligned}
\alpha &= \frac{k}{\sqrt{2}} \left[1 + \sqrt{1 + \frac{4m\omega^2}{EIk^4}} \right]^{1/2} \\
\beta &= \frac{k}{\sqrt{2}} \left[-1 + \sqrt{1 + \frac{4m\omega^2}{EIk^4}} \right]^{1/2}
\end{aligned}
\right\}
\tag{9}
$$

Use of the boundary conditions, Eqs. (3), results in the following characteristic equation

$$
\begin{vmatrix}
0 & 1 & 0 & 1 \\
\alpha & 0 & \beta & 0 \\
-\alpha^2 \sin \alpha L & -\alpha^2 \cos \alpha L & \beta^2 \sinh \beta L & \beta^2 \cosh \beta L \\
-\alpha^3 \cos \alpha L & \alpha^3 \sin \alpha L & \beta^3 \cosh \beta L & \beta^3 \sinh \beta L
\end{vmatrix} = 0
\tag{10}
$$

The expansion of the determinant is:

$$(\beta \sinh \beta L + \alpha \sin \alpha L)(\beta^3 \sinh \beta L - \alpha^3 \sin \alpha L)$$
$$= (\beta^2 \cosh \beta L + \alpha^2 \cos \alpha L)^2 \qquad (11)$$

After performing the indicated operations, we obtain

$$-(\alpha^4 + \beta^4) - 2\alpha^2\beta^2 \cos \alpha L \cosh \beta L + \alpha\beta(\beta^2 - \alpha^2) \sin \alpha L \sinh \beta L = 0 \qquad (12)$$

From Eqs. (9) we have that

$$\left. \begin{array}{l} \beta^2 - \alpha^2 = -k^2 \\[2mm] \alpha\beta = \left(\dfrac{m\omega^2}{EI}\right)^{1/2} \\[4mm] (\alpha^4 + \beta^4) = k^4\left[1 + \dfrac{2m\omega^2}{EI}\right] \end{array} \right\} \qquad (13)$$

Sutstitution of the above expressions into Eq. (12) yields

$$k^4 + k^2\left(\frac{m\omega^2}{EI}\right)^{1/2} \sin \alpha L \sinh \beta L + \frac{2m\omega^2}{EI} + 2\frac{m\omega^2}{EI} \cos \alpha L \cosh \beta L = 0 \qquad (14)$$

Let

$$\lambda = \frac{\bar{P}}{P_E} \quad \text{and} \quad \bar{\omega} = \omega\left(\frac{m}{EI}\right)^{1/2}\left(\frac{L}{\pi}\right)^2 \qquad (15)$$

Then Eq. (14) becomes

$$\lambda^2 + \lambda\bar{\omega} \sin \alpha L \sinh \beta L + 2\omega^{-2} + 2\omega^{-2} \cos \alpha L \cosh \beta L = 0 \qquad (16)$$

where

$$\left. \begin{array}{l} \alpha L = \pi\left(\dfrac{\lambda}{2}\right)^{1/2}\left[1 + \sqrt{1 + 4\dfrac{\omega^{-2}}{\lambda^2}}\right]^{1/2} \\[5mm] \beta L = \pi\left(\dfrac{\lambda}{2}\right)^{1/2}\left[-1 + \sqrt{1 + 4\dfrac{\omega^{-2}}{\lambda^2}}\right]^{1/2} \end{array} \right\} \qquad (17)$$

If we solve the transcendental equation, Eq. 16, for $\bar{\omega}^2$ by assigning different positive values for λ, starting from zero, we obtain the plot in Fig. 8-5 (qualitative). This plot clearly shows that as long as $\lambda < 2.01$, the motion is oscillatory, but for $\lambda > 2.01$, $\bar{\omega}$ becomes complex and the motion becomes divergent. Therefore, $\lambda_{cr} = 2.01$.

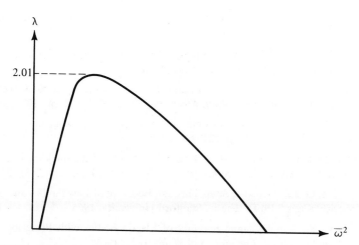

Figure 8-5. Critical condition for the cantilever under a follower force.

8.3 BUCKLING UNDER DYNAMIC LOADS

There have been numerous studies in recent years on buckling under dynamic loads. When the load is explicitly dependent on time, the system is nonconservative, and the only approach that should be used in such cases is the dynamic approach. Although the idea of dynamic stability is conceptually simple, its quantitative application to problems of elastic structural elements and configurations is, in many cases, extremely difficult. References 5 and 6 deal with a number of dynamic stability problems, definitions of critical dynamic loads, and approaches to establish these critical conditions. Dynamic loads that are conservative have received wide attention. Loads of constant magnitude and infinite duration, and impulsive loads, fall in this category. In addition to some of the papers in Ref. 6, interesting and valuable information is presented in the works of Hoff and Bruce (Ref. 9), Simitses (Ref. 10) Hsu (Refs. 11 and 12) and Lock (Ref. 13).

REFERENCES

1. ZIEGLER, H., *Principles of Structural Stability*, Blaisdell Publishing Co., Waltham, Massachusetts, 1968.

2. LEIPHOLZ, H., *Stability Theory*, Academic Press, New York, 1970.

3. BOLOTIN, V. V., *Nonconservative Problems of the Theory of Elastic Stability*, edited by G. Herrmann (translated from the Russian), The Macmillan Co., New York, 1963.

4. HERRMANN, G., "Dynamics and Stability of Mechanical Systems with Follower Forces," NASA Contractor Report CR-1782, Nov, 1971.

5. BOLOTIN, V. V., *Dynamic Stability of Elastic Systems*, Holden-Day Inc., San Francisco, 1964.

6. HERRMAN, G., ed., "Dynamic Stability of Structures," *Proc. Int. Conf.*, October, 1965, Pergamon Press Inc., New York, 1967.

7. HERRMANN, G., "Stability of Equilibrium of Elastic Systems Subjected to Nonconservative Forces," *Appl. Mech. Rev.*, Vol. 20, No. 2, pp. 103–108, 1967.

8. BECK, M., "Die Knicklast des einseitig eingespannten, tangential gedrückten Stabes," *ZAMP*, Vol. 3, pp. 225–228, 1952.

9. HOFF, N. J., and BRUCE, V. G., "Dynamic Analysis of the Buckling of Laterally Loaded Flat Arches," *J. Math. and Phys.*, Vol. XXXII, pp. 276–288, 1954.

10. SIMITSES, G. J., "Dynamic Snap-Through Buckling of Law Arches and Shallow Spherical Caps," Ph.D. Thesis, Stanford University, 1965.

11. HSU, C. S., "On Dynamic Stability of Elastic Bodies with Prescribed Initial Conditions," *Int'l J. Eng. Sci.*, Vol. 4, pp. 1–21, 1966.

12. HSU, C. S., "The Effects of Various Parameters on the Dynamic Stability of a Shallow Arch," *J. Appl. Mech.*, Vol. 34, No. 2, pp. 349–358, 1967.

13. LOCK, M. H., "The Snapping of a Shallow Sinusoidal Arch under a Step Pressure Load," *AIAA J.*, Vol. 4, No. 7, pp. 1249–1256, 1966.

APPENDIX

WORK AND ENERGY RELATED
PRINCIPLES AND THEOREMS

This appendix will summarize the work and energy principles (and theorems derived from these) that have been used in stability analysis and are directly referred to in this text. In addition, some explanations and definitions will be given to facilitate understanding and application of these concepts. Because of this, only the principles and derived theorems associated with virtual work will be treated. Complementary energy and complementary virtual work concepts, principles, and theorems are not included. The student interested in an extensive and thorough treatise of all the work and energy principles is referred to the texts of Langhaar (Ref. 1), Argyris (Ref. 2), Dym and Shames (Ref. 3), and Fung (Ref. 4). One of the first and best-written texts on the subject (with numerous applications on structural problems) is the book by Hoff (Ref. 5).

A.1 STRAIN ENERGY

A deformable body is said to be perfectly elastic if the state of stress and the corresponding state of strain are the same for the same level of the external forces regardless of the order of application of the loads and of whether this level is during loading or unloading of some or all of the loads. This statement is clearly understandable when related to the simple tensile

227

test. If the stress-strain relation for such a test is the same during the loading and unloading processes, the behavior is called elastic and the specimen is called a perfectly elastic body.

If a perfectly elastic body is under the action of external loads (distributed and concentrated forces, distributed and concentrated moments), the body deforms and work is done by these external loads. This work, in the absence of kinetic energy (quasistatic application of the loads), is stored in the system. Because of the assumption that the material is perfectly elastic, the work done by the loads can be regained if the loads are quasistatically decreased to zero. The energy stored in the system is known as the strain energy.

If we consider a deformable body at state I and apply a set of loads that strain the body to state II, and if we use a cartesian reference frame, x, y, z, the work done by these forces, W_e, is equal to the strain energy, U_i, and it is given by (for small strains)

$$W_e = U_i = \int_V \left\{ \int_I^{II} \left[\tau_{xx}\, d\epsilon_{xx} + \tau_{yy}\, d\epsilon_{yy} + \tau_{zz}\, d\epsilon_{zz} \right. \right.$$
$$\left. \left. + \tau_{xy}\, d\gamma_{xy} + \tau_{yz}\, d\gamma_{yz} + \tau_{zx}\, d\gamma_{zx} \right] \right\} dV \qquad \text{(A-1a)}$$

or

$$U_i = \int_V \bar{U}_i\, dV \qquad \text{(A-1b)}$$

where $\bar{U}_i$ is defined as the strain-energy density (strain energy per unit volume).

The existence of the strain-energy density function and the energy balance expressed by Eqs. (A-1) is in agreement with the first and second laws of thermodynamics for isentropic processes. In this case, the energy stored in the system is called internal energy. In addition, if the process is a reversible isothermal one, then the stored energy is often called the free energy (see Refs. 3 and 4). In effect, the strain-energy density represents the energy that can be converted to mechanical work in a reversible adiabatic or isothermal process.

Since the strain-energy density at a point depends on the state of strain, the incremental strain-energy density, which is a perfect differential for perfectly elastic behavior, $d\bar{U}_i$, is given by

$$d\bar{U}_i = \frac{\partial \bar{U}_i}{\partial \epsilon_{xx}}\, d\epsilon_{xx} + \frac{\partial \bar{U}_i}{\partial \epsilon_{yy}}\, d\epsilon_{yy} + \frac{\partial \bar{U}_i}{\partial \epsilon_{zz}}\, d\epsilon_{zz}$$
$$+ \frac{\partial \bar{U}_i}{\partial \gamma_{xy}}\, d\gamma_{xy} + \frac{\partial \bar{U}_i}{\partial \gamma_{yz}}\, d\gamma_{yz} + \frac{\partial \bar{U}_i}{\partial \gamma_{zx}}\, d\gamma_{zx} \qquad \text{(A-2)}$$

From Eqs. (A-1), it can be seen that

$$\tau_{xx} = \frac{\partial \bar{U}_i}{\partial \epsilon_{xx}}, \qquad \tau_{yy} = \frac{\partial \bar{U}_i}{\partial \epsilon_{yy}}, \qquad \tau_{zz} = \frac{\partial \bar{U}_i}{\partial \epsilon_{zz}}$$

$$\tau_{xy} = \frac{\partial \bar{U}_i}{\partial \gamma_{xy}}, \qquad \tau_{yz} = \frac{\partial \bar{U}_i}{\partial \gamma_{yz}}, \qquad \tau_{zx} = \frac{\partial \bar{U}_i}{\partial \gamma_{zx}}$$

(A-3)

When the material follows Hooke's law (linearly elastic behavior), then

$$\bar{U}_i = \tfrac{1}{2}[\tau_{xx}\epsilon_{xx} + \tau_{yy}\epsilon_{yy} + \tau_{zz}\epsilon_{zz} + \tau_{xy}\gamma_{xy} + \tau_{yz}\gamma_{yz} + \tau_{zx}\gamma_{zx}] \qquad (A\text{-}4)$$

If the linear stress-strain relations are used in Eq. (A-4) in terms of Poisson's ratio, v, and Young's modulus of elasticity, E, the strain-energy density can be expressed solely either in terms of strains or in terms of stresses.

Three-dimensional Case

$$\bar{U}_i = \frac{E}{2(1+v)(1-2v)}\Big[(1-v)(\epsilon_{xx}^2 + \epsilon_{yy}^2 + \epsilon_{zz}^2) + 2v(\epsilon_{xx}\epsilon_{yy}$$

$$+ \epsilon_{yy}\epsilon_{zz} + \epsilon_{zz}\epsilon_{xx}) + \frac{1-2v}{2}(\gamma_{xy}^2 + \gamma_{yz}^2 + \gamma_{zx}^2)\Big] \qquad (A\text{-}5a)$$

$$\bar{U}_i = \frac{1}{2E}[(\tau_{xx}^2 + \tau_{yy}^2 + \tau_{zz}^2) - 2v(\tau_{xx}\tau_{yy} + \tau_{yy}\tau_{zz}$$

$$+ \tau_{yy}\tau_{xx}) + 2(1+v)(\tau_{xy}^2 + \tau_{yz}^2 + \tau_{zx}^2)] \qquad (A\text{-}5b)$$

Two-dimensional Case

1. Plane stress (xy-plane):

$$\bar{U}_i = \frac{1}{2(1-v^2)}\Big[\epsilon_{xx}^2 + \epsilon_{yy}^2 + 2v\epsilon_{xx}\epsilon_{yy} + \frac{1-v}{2}\gamma_{xy}^2\Big] \qquad (A\text{-}6a)$$

$$\bar{U}_i = \frac{1}{2E}[\tau_{xx}^2 + \tau_{yy}^2 - 2v\tau_{xx}\tau_{yy} + 2(1+v)\tau_{xy}^2] \qquad (A\text{-}6b)$$

2. Plane strain (xy-plane):

$$\bar{U}_i = \frac{E}{2(1+v)(1-2v)}\Big[(1-v)(\epsilon_{xx}^2 + \epsilon_{yy}^2) + 2v\epsilon_{xx}\epsilon_{yy} + \frac{1-2v}{2}\gamma_{xy}^2\Big]$$

(A-7a)

$$\bar{U}_i = \frac{1}{2E}[(1-v^2)(\tau_{xx}^2 + \tau_{yy}^2) - 2v(1+v)\tau_{xx}\tau_{yy} + 2(1+v)\tau_{xy}^2] \qquad (A\text{-}7b)$$

One-dimensional Case. For this case, let us consider a Euler-Bernoulli beam with the xz-plane as a plane of structural symmetry. Then $\epsilon_{zz} = \gamma_{xz} = 0$ and

$$\vec{U}_i = \frac{E}{2}\epsilon_{xx}^2 \tag{A-8a}$$

$$\bar{U}_i = \frac{1}{2E}\tau_{xx}^2 \tag{A-8b}$$

A.2 THE PRINCIPLE OF VIRTUAL DISPLACEMENTS OR VIRTUAL WORK

Before we state this principle, we must first clearly explain what is meant by "virtual displacement." A virtual displacement is a hypothetical displacement which must be compatible with the constraints for a given problem. If we deal with a particle, a virtual displacement is a single vector without any limitations on magnitude and direction. If we deal with a *rigid* body, a virtual displacement is a displacement field $\bar{u}(x, y, z)$ which must be compatible with the requirement that the body be rigid (see Fig. A-1). Note that in Fig. A-1a, in addition to the rotation θ, the system may be translated, and the combination comprises a virtual displacement. Lastly, when we deal with a cohesive deformable continuum the virtual displacement must be compatible with (1) the constitution of the medium, and (2) the associated method of analysis. The latter statement means that the virtual displacements must be consistent with the theory and its related kinematic assumptions that lead to the field equations that govern the response of the system to any set of external causes.

First, what we mean by compatible with the constitution of the medium is that, since we deal with a cohesive continuum, the virtual displacement components must be single-valued continuous functions of positions (material points coordinates) with continuous derivatives. Second, compatible with the associated method of analysis implies the following: (1) Since we are interested in deformations in the analysis of deformable bodies, then

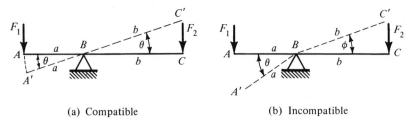

(a) Compatible   (b) Incompatible

Figure A-1. Compatible and incompatible virtual displacements for a rigid bar.

the virtual displacement is a deformation field $\bar{u}(x, y, z)$ which is consistent with the kinematic constraints on the bounding surface. (2) Since there are different approximating theories describing the kinematics of the problem, as for example the theory of small deformation gradient, the virtual deformation field must be consistent with these approximations. Therefore, a deformation field which has these properties is referred to as a kinematically admissible field, and thus any kinematically admissible field can be used as a virtual displacement. Finally, the reason the virtual displacement is called hypothetical is that during virtual displacements the forces, internal and external, are kept constant, which is not compatible with the behavioral response of systems, in general.

The principle of virtual displacements or virtual work may be stated as follows:

> A body is in equilibrium, under a given system of loads, if and only if for any virtual displacement the work done by the external forces is equal to the strain energy.

Note that:

1. A principle in mechanics is like an axiom in mathematics. There is no proof of a principle, although one may show its equivalence to another principle or law.

2. If we realize that a virtual displacement is kinematically admissible and that the forces are kept constant during virtual displacements, the principle holds for deformable bodies as well as rigid bodies and particles. In the case of rigid bodies and particles, the strain energy is zero.

3. The mathematical expression for the principle is

$$\delta_{\cdot\cdot_\epsilon\cdot\cdot}W = \delta_{\cdot\cdot_\epsilon\cdot\cdot}U_i \qquad \text{(A-9)}$$

for deformable bodies, and

$$\delta_{\cdot\cdot_\epsilon\cdot\cdot}W = 0 \qquad \text{(A-10)}$$

for rigid bodies and particles. In Eqs. (A-9) and (A-10), $\delta_{\cdot\cdot_\epsilon\cdot\cdot}W$ represents the work done by the external forces and $\delta_{\cdot\cdot_\epsilon\cdot\cdot}U_i$ is the strain energy during a virtual displacement denoted by the subscript "ϵ".

A few simple applications of the principle are given below.

1. A Particle under N Forces. Given a particle under the application of N forces $\vec{F}_i$, according to the principle, this particle is in equilibrium if and only if

$$\delta_{\cdot\cdot_\epsilon\cdot\cdot}W = 0$$

Let $\bar{u}$ be a virtual displacement. Then by the principle

$$\vec{F}_1 \cdot \vec{u} + \vec{F}_2 \cdot \vec{u} + \cdots + \vec{F}_u \cdot \vec{u} = 0$$

or

$$(\bar{F}_1 + \bar{F}_2 + \bar{F}_3 + \cdots + \bar{F}_n) \cdot \vec{u} = 0$$

$$\left(\sum_{i=1}^{N} \vec{F}_i \right) \cdot \vec{u} = 0 \qquad (A\text{-}11)$$

For this to be true, either $\sum_{i=1}^{u} \vec{F}$ is normal to $\vec{u}$ or zero. But since $\vec{u}$ is any displacement vector, then $\sum_{i=1}^{N} F_i$ must be zero. This is in complete agreement with the necessary and sufficient conditions for equilibrium of a particle under static loads which are derived from Newton's second law.

2. The Fulcrum Problem. Consider the rigid bar of Fig. A-2

The virtual displacement consists of a translation in the positive y-direction and a rotation θ as shown. (Note that the rigid bar ACB which is originally straight remains straight, $A'C'B'$, during the virtual displacement (compatible with the fact that the bar is rigid).

The work done by the forces during the virtual displacement is zero

$$(-F_1\vec{j}) \cdot \overrightarrow{AA'} + (R\vec{j}) \cdot \overrightarrow{CC'} + (-F_2\vec{j}) \cdot \overrightarrow{BB'} = 0 \qquad (A\text{-}12)$$

where $\overrightarrow{AA'}$, $\overrightarrow{CC'}$, and $\overrightarrow{BB'}$ are position vectors from A to A', C to C', and B to B', respectively

$$\overrightarrow{AA} = a(1 - \cos\theta)\vec{i} + (d - a\sin\theta)\vec{j}$$
$$\overrightarrow{CC'} = d\vec{j} \qquad (A\text{-}13)$$
$$\overrightarrow{BB'} = -b(1 - \cos\theta)\vec{i} + (d + b\sin\theta)\vec{j}$$

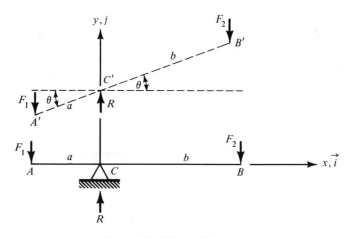

Figure A-2. Fulcrum Geometry.

Substitution of Eqs. (A-13) into Eq. (A-12) yields

$$-E_1(d - a \sin \theta) + Rd - F_2(d + b \sin \theta) = 0$$

or

$$(-F_1 + R - F_2)d + (F_1 a - F_2 b) \sin \theta = 0 \qquad \text{(A-14)}$$

Since d and θ are independent (we can have $d \neq 0$ and $\theta \equiv 0$ or $\theta \neq 0$ and $d \equiv 0$), then

$$F_1 + F_2 = R \quad \text{and} \quad F_1 a = F_2 b \qquad \text{(A-15)}$$

These equations are in complete agreement with the necessary and sufficient conditions for equilibrium of a rigid body (sum of forces equals zero, and sum of moments about C equals zero, respectively).

3. Extension of a Bar.   Consider the straight bar shown in Fig. A-3. Making the usual linear theory assumptions (small deformation gradients and linearly elastic behavior) and reducing the problem to a one-dimensional one, we may write

$$\epsilon_{xx} = \frac{du}{dx}$$
$$\tau_{xx} = E\epsilon_{xx} \qquad \text{(A-16)}$$

where u is a function of x only. If we allow $\bar{u}$ to denote a virtual displacement, then $\bar{u}$ must be kinematically admissible and $\bar{u}(0) = 0$. On this basis, the corresponding virtual strain is given by

$$\bar{\epsilon}_{xx} = \frac{d\bar{u}}{dx} \qquad \text{(A-17)}$$

The virtual work and corresponding strain energy are given by

$$\delta_\epsilon W = P\bar{u}(L) \qquad \text{(A-18a)}$$

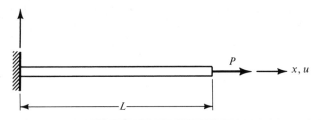

Figure A-3. Bar Geometry.

and

$$\delta_\epsilon U_i = \int_V \tau_{xx} \bar{\epsilon}_{xx} \, dV$$

$$= \int_0^L (\tau_{xx} A) \frac{d\bar{u}}{dx} \, dx \qquad \text{(A-18b)}$$

Integration by parts yields

$$\delta_\epsilon \bar{U}_i = (\tau_{xx} A \bar{u}) \Big|_0^L - \int_0^L \frac{d}{dx} (\tau_{xx} A) \bar{u} \, dx \qquad \text{(A-19)}$$

Since $\bar{u}(0) = 0$, by the principle we obtain

$$[(P - A\tau_{xx}) \bar{u}]_{x=L} - \int_0^L \frac{d}{dx} (\tau_{xx} A) \bar{u} \, dx = 0 \qquad \text{(A-20)}$$

Because $\bar{u}$ is a virtual displacement, then

$$\frac{d}{dx} (\tau_{xx} A) = 0 \longrightarrow \tau_{xx} A = \text{constant}$$

and $P = A \tau_{xx}$ since $\bar{u}(L)$ is arbitrary. The system is in equilibrium if and only if $\tau_{xx} = P/A = \text{constant}$.

A.3   DERIVATIVES OF THE PRINCIPLE OF VIRTUAL WORK

A number of principles, theorems, and methods may be considered as direct derivatives from the principle of virtual work. The most pertinent of these derivatives are listed in this section.

A.3-1   The Principle of the Stationary Value of the Total Potential

If the system is conservative, then the work done by the loads from the zero deformation state (strain-free position) to a final state is equal to the negative change in the total potential of the external forces, U_p. If this potential is defined such that it is zero at the zero deformation state, then

$$-W = U_p \qquad \text{(A-21)}$$

Next, the variation in the work done during virtual displacement is related to the variation in the potential of the external forces by

$$\delta_\epsilon W = -\delta_\epsilon U_p \qquad \text{(A-22)}$$

Substitution of Eq. (A-22) into Eq. (A-9) yields

$$\delta_\epsilon U_i + \delta_\epsilon U_p = 0$$

or

$$\delta_\epsilon(U_i + U_p) = \delta_\epsilon U_T = 0 \qquad \text{(A-23)}$$

where U_T is called the total potential (energy) of the system. This equation implies that:

> An elastic deformable system is in equilibrium (static) if and only if the first variation of the total potential vanishes for every virtual displacement.

Note that virtual displacements must be kinematically admissible and that the loads and stresses remain constant during such deformations.

If we consider that the total potential is a function of N deformation parameters q_i, and that a virtual displacement can be taken to be any one of these q_i's, then by Eq. (A-23)

$$\delta_\epsilon U_T = \frac{\partial U_T}{\partial q_i} \delta q_i \qquad i = 1, 2, \ldots, N \qquad \text{(A-24)}$$

Since δq_i is arbitrary, this equation implies that

$$\frac{\partial U_T}{\partial q_i} = 0 \qquad i = 1, 2, \ldots, N \qquad \text{(A-25)}$$

for static equilibrium.

By Eq. (A-25) the total differential of U_T must be zero, or

$$dU_T = \sum_{i=1}^{N} \frac{\partial U_T}{\partial q_i} dq_i = 0 \qquad \text{(A-26)}$$

This argument may be extended to a deformable system, and we conclude that the vanishing of δU_T implies the venishing of dU_T. Next, since dU_T vanishes at stationary points (relative minima, maxima, or saddle points), and U_T is said to have a stationary value at such points, then Eq. (A-23) may be interpreted as the mathematical expression of the following principle.

> An elastic deformable system is in equilibrium (static) if and only if the total potential has a stationary value.

An equivalent statement to the above is:

> Of all possible kinematically admissible deformation fields in an elastic conservative system, for a specified level of the external loads and the corresponding internal loads, only those corresponding to equilibrium (static) make the total potential assume a stationary value.

This statement is known as the *principle of the stationary value of the total potential.* In reality, it is a theorem because it is derived from and proven by a basic principle, the *principle of virtual work.*

A.3-2   The Principle of the Minimum Total Potential

The above theorem is easily extended to an equivalent of the Lagrange-Dirichlet theorem (see Chapter 1) for an elastic conservative system by requiring the stationary value to be a relative minimum. If this happens, the equilibrium is stable. Often, this theorem is referred to as the *principle of the minimum total potential* and it is given below.

> Of all possible kinematically admissible deformation fields in an elastic conservative system, for a specified level of the external loads and the corresponding internal loads, only those that make the total potential assume a minimum value correspond to a stable equilibrium.

A.3-3   Castigliano's First Theorem (Part I)

Another derivative of the principle of virtual work is Castigliano's first theorem, part I.

Consider an elastic system under the action of N concentrated loads, P_j, (forces and moments). Let y_j denote the components of deformation (or rotations) at the points of applications of the forces (or moments) and in the directions of these loads. If δy_j denote virtual displacements, then the virtual work is given by

$$\delta W = \sum_{j=1}^{N} P_j \delta y_j \qquad \text{(A-27)}$$

If we can express the deformation components of the material points on the body in terms of the y_j components, the stresses, strains, and consequently the strain energy of the elastic system become functions of the y_j components, the structural geometry, and the elastic behavior (stress-strain law which need not necessarily be linearly elastic). If we now give each component y_j a small variation δy_j (virtual displacement), then

$$\delta U_i = \sum_{j=1}^{N} \frac{\partial U_i}{\partial y_j} \delta y_j \qquad \text{(A-27)}$$

By the principle of virtual work

$$\sum_{j=1}^{N} \left(\frac{\partial U_i}{\partial y_j} - P_j \right) \delta y_j = 0 \qquad \text{(A-28)}$$

Therefore, since the virtual displacements are independent, we have the mathematical expression of Castigliano's first theorem

$$\frac{\partial U_i}{\partial y_j} = P_j \tag{A-29}$$

Note that this theorem applies to elastic systems regardless of the behavior (nonlinear elasticity behavior as well). For applications, see Refs. 6 and 7.

One important application of the theorem is in finding reaction forces for structural systems. For example, if the deformation component is known to be zero at some point and for a given direction, first we let y_r exist; then we express U_i in terms of y_r. Finally, the sought reaction is equal to $(\partial U_i/\delta y_r)$, evaluated at $y_r = 0$, according to Eq. (A-29).

A.3-4   The Unit-Displacement Theorem

Another important derivative of the principle of virtual work is the unit-displacement theorem. This theorem is used to determine the load P_r (force or moment) necessary to maintain equilibrium in an elastic system when the distribution of true stresses is known. Let the true stresses be given by $(\tau_{xx}, \tau_{yy}, \ldots, \tau_{zx})$. Consider a virtual displacement δy_r at the point of application and in the direction of P_r. This virtual displacement produces virtual strains $\delta\epsilon_{ij}^r$ and according to the principle of virtual work

$$P_r\delta y_r = \int_V (\tau_{xx}\delta\epsilon_{xx}^r + \tau_{yy}\delta\epsilon_{yy}^r + \cdots + \tau_{zx}\delta_{zx}^r)\, dV \tag{A-30}$$

In a linearly elastic system, the virtual strains $\delta\epsilon_{ij}^r$ are proportional to y_r:

$$\delta\epsilon^r = \epsilon^r\delta y_r \tag{A-31}$$

where ϵ_r represents compatible strains due to a unit virtual displacement ($\delta y_r = 1$). Assuming, therefore, that $\delta y_r = 1$, Eq. (A-30) becomes

$$P_r = \int_V (\tau_{xx}\epsilon_{xx}^r + \tau_{yy}\epsilon_{yy}^r + \cdots + \tau_{zx}\gamma_{zx}^r)\, dV \tag{A-32}$$

This equation is the mathematical expression of the unit-displacement theorem, which is stated below:

> The force necessary to maintain equilibrium under a specified stress distribution (which is derived from a specified deformation state) is given by the integral over the volume of true stresses τ_{ij} multiplied by strains ϵ_{ij}^r compatible with a unit displacement at the point and in the direction of the required force.

This theorem, because of Eq. (A-31), is restricted to a system with linearly elastic behavior. For a move extensive treatment and applications, see Refs. 5, 6, 7, and 8.

Some authors refer to the above as the unit-dummy-displacement method (not a theorem). This method or theorem may be used very effectively for the calculation of stiffness properties of structural elements employed in matrix methods of structural analysis (see Refs. 6 and 8).

A.3-5   The Rayleigh-Ritz Method

A variational formulation of a boundary-value problem is very useful for the approximate computation of the solution. One of the most widely used approximate methods is the Rayleigh-Ritz or simply Ritz method. This method was first employed by Lord Rayleigh (Ref. 9) in studies of vibrations and by Timoshenko (Ref. 10) in buckling problems. The method was refined and extended by Ritz (Ref. 11), and since then it has been applied to numerous problems in applied mechanics including deformation analyses, stability, and vibrations of complex systems. Although the method is based on the variational formulation of a specific problem, it may be considered as a derivative of the principle of the stationary value of the total potential when applied to elastic systems under quasistatic loads.

The basic ideas of the method are outlined by using as an example the deformation analysis of a general three-dimensional elastic system under the application of quasistatic loads (stable equilibrium). For a more rigorous treatment of the method from a mathematical (variational) point of view, refer to the texts of Courant and Hilbert (Ref. 12), Gelfand and Fomin (Ref. 13), and Kantorovich and Krylov (Ref. 14).

An elastic system consists of infinitely many material points; consequently, it has infinitely many degrees of freedom. By making certain assumptions about the nature of the deformations, we can reduce the elastic system to one with finite degrees of freedom. For instance, the deformation components u, v, and w may be represented by a finite series of kinematically admissible functions multiplied by undetermined constants

$$u(x, y, z) = \sum_{i=1}^{N} \bar{u}_i(x, y, z) = \sum_{i=1}^{N} a_i f_i(x, y, z)$$

$$v(x, y, z) = \sum_{i=1}^{N} \bar{v}_i(x, y, z) = \sum_{i=1}^{N} b_i g_i(x, y, z) \qquad \text{(A-33)}$$

$$w(x, y, z) = \sum_{i=1}^{N} \bar{w}_i(x, y, z) = \sum_{i=1}^{N} c_i h_i(x, y, z)$$

Note that, if we use small-deformation gradient theory, what is meant by kinematic admissibility is that the functions f_i, g_i, and h_i must be single-valued, continuous, differentiable, and must satisfy the kinematic boundary conditions. Then with Eqs. (A-33) the total potential, which is a functional, becomes a function of the three-N undetermined constants a_i, b_i, and c_i or

$$U_T[u, v, w] = U_T(a_i, b_i, c_i) \qquad \text{(A-34)}$$

Now, since the functions f_i, g_i, and h_i are kinematically admissible, the virtual displacements can be taken as

$$\delta u = \delta a_i f_i, \qquad \delta v = \delta b_i g_i, \qquad \delta w = \delta c_i h_i \qquad \text{(A-35)}$$

and the variation in the total potential is given by

$$\delta U_T = \sum_{i=1}^{N} \left[\frac{\partial U_T}{\partial a_i} \delta a_i + \frac{\partial U_T}{\partial b_i} \delta b_i + \frac{\partial U_T}{\partial c_i} \delta c_i \right] \qquad \text{(A-36)}$$

Therefore, the elastic system is in equilibrium if

$$\frac{\partial U_T}{\partial a_i} = 0, \qquad \frac{\partial U_T}{\partial b_i} = 0, \qquad \frac{\partial U_T}{\partial c_i} = 0 \qquad i = 1, 2, \ldots, N \qquad \text{(A-37)}$$

Equations (A-37) represent a system of three-N linearly independent algebraic equations in the *three-N* undetermined constants a_i, b_i, and c_i. The solution of these systems yields the values for these constants, and substitution into Eqs. (A-33) leads to the approximate expressions for the deformation components u, v, and w. Once these are known, we can evaluate the strains from the kinematic relations, and consequently the stresses from the constitutive relations. Thus the analysis is complete because we know the state of deformation and the stress and strain at every material point.

A number of questions arise, as far as the method is concerned, regarding the choice of the functions f_i, g_i, and h_i and the accuracy of the solution (convergence). These questions are discussed rigorously and in detail in Refs. 12–15. In summary, some of the important conclusions, in answer to these questions, are:

1. The Rayleigh-Ritz method is applicable to variational problems which satisfy the sufficiency conditions for a minimum (maximum) of a functional. The central idea is that of a minimizing (maximizing) sequence. A sequence $\bar{u}_1, \bar{u}_2, \ldots, \bar{u}_N$ (and consequently $\bar{v}_1, \bar{v}_2, \ldots, \bar{v}_N$, and $\bar{w}_1, \bar{w}_2, \ldots, \bar{w}_N$) of kinematically admissible functions is called a minimizing (maximizing) sequence if

$$U_T \left[\sum_{i=1}^{N} \bar{u}_i, \sum_{i=1}^{N} \bar{v}_i, \sum_{i=1}^{N} \bar{w}_i \right]$$

converges to the minimum (maximum) of $U_T[u, v, w]$.

2. A minimizing (maximizing) sequence converges to a minimizing

(maximizing) function, if constructed properly, for all one-variable problems (beams, columns) and all two- and three-variable problems (plates and shells) in which the order of the Euler differential equation is at least four.

3. A properly constructed minimizing (maximizing) sequence must be complete. This means that we select a set $\bar{u}_1, \bar{u}_2, \ldots, \bar{u}_N$ of admissible functions such that any admissible function [including the minimizing (maximizing) function] and its derivatives can be approximated arbitrarily closely by a suitable linear combination

$$a_1\bar{u}_1 + a_2\bar{u}_2 + \cdots + a_N\bar{u}_N \qquad \text{(A-38)}$$

For example, in one-dimensional problems, $(L - x)x^{n+1}$ $(n = 0, 1, 2, \ldots)$ is a complete set vanishing on the boundary of the interval $0 \leq x \leq L$. Similarly, $\sin(n\pi x/L)$ $(n = 1, 2, \ldots)$ for the same case. Again in one-dimensional problems, $x^2(L - x)^{n+2}$ $(n = 0, 1, 2, \ldots)$ is a complete set vanishing on the boundary, along with its first derivative, of the interval $0 \leq x \leq L$. Similarly, $\cos(n\pi x/L) - \cos[(n + 2)\pi x/L]$ $(n = 0, 1, 2, \ldots)$ is a complete set for this latter case.

4. When the Rayleigh-Ritz method is used for beam, plate, and shell problems, it leads to fairly accurate expressions for the deformations. If one is interested in rotations, moments, and transverse shears, the accuracy decreases, respectively, because these quantities expressed in terms of deformations require higher derivatives of the deformations, and the derivatives are less accurate approximations than the functions themselves.

5. Also, because of the reasons given in item 4, equilibrium at a point is not satisfied exactly. Stresses computed through approximate deformations do not, in general, satisfy equilibrium equations.

The application of the Rayleigh-Ritz to stability problems is presented in Chapter 5 of this text. As an application of the method for stable equilibrium, consider the beam shown in Fig. A-4. Using pure bending theory, the

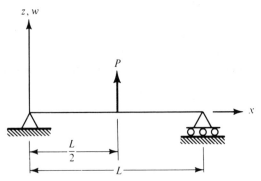

Figure A-4. Beam Geometry.

total potential is given by

$$U_T = \frac{EI}{2} \int_0^L (w'')^2 \, dx - Pw\left(\frac{L}{2}\right) \qquad \text{(A-38)}$$

Since $\sin(m\pi x/L)$ $(m = 1, 2, \ldots, N)$ satisfy the kinematic boundary conditions, let

$$w = \sum_{m=1}^N a_m \sin \frac{m\pi x}{L} \qquad \text{(A-39)}$$

Substitution of Eq. (A-39) into Eq. (A-38) yields

$$U_T = \frac{EIL}{4} \sum_{m=1}^N a_m^2 \left(\frac{m\pi}{L}\right)^4 - P \sum_{m=1}^N a_m \sin \frac{m\pi}{2} \qquad \text{(A-40)}$$

By the principle of the minimum of the total potential,

$$\frac{EIL}{2} a_m \left(\frac{m\pi}{L}\right)^4 - P \sin \left(\frac{m\pi}{2}\right) = 0 \qquad m = 1, 2, \ldots, N \qquad \text{(A-41)}$$

Equations (A-41) represent a decoupled system of N linear algebraic equations in a_m $(m = 1, 2, \ldots, N)$. The solution is

$$a_m = \frac{2P}{LEI}\left(\frac{L}{m\pi}\right)^4 \sin \frac{m\pi}{2}$$

and

$$w = \sum_{n=1}^N \frac{2P}{LEI}\left(\frac{L}{m\pi}\right)^4 \sin \frac{m\pi}{2} \sin \frac{m\pi x}{L} \qquad \text{(A-42)}$$

From, Table A-1, we see that the convergence is very rapid. Although the convergence for the deformation is very rapid, this is not so for the moment and shear. For example, the shear, $V(x)$, is given by

$$V(x) = -EIw''' = \frac{2P}{\pi} \sum_{m=1}^N \frac{1}{m} \sin \frac{m\pi}{2} \cos \frac{m\pi x}{L} \qquad \text{(A-43)}$$

Table A-1. COMPARISON WITH THE EXACT SOLUTION

Solution	$\frac{EI}{PL^3} w$ at			
	$L/8$	$L/4$	$3L/8$	$L/2$
Exact	0.00765	0.01432	0.01904	0.02083
One-term	0.00806	0.01452	0.01891	0.02053
Two-term	0.00786	0.01432	0.01906	0.02081
Three-term	0.00765	0.01432	0.01904	0.02083

In addition, because the minimizing sequence is orthogonal, N can be taken as infinity, in which case

$$w(x) = \frac{2PL^3}{EI\pi^4} \sum_{m=1}^{\infty} \frac{1}{m^4} \sin \frac{m\pi}{2} \sin \frac{m\pi x}{L} \qquad \text{(A-44)}$$

In such cases, if the series can be closed or if all of the terms are considered, the Rayleigh-Ritz method gives exact results. The series may be closed when evaluated at a point or in general (for any x) through different mathematical operations such as the calculus of residues (Ref. 16), integration of series (Refs. 17 and 18), and others. (See Table A-2 for typical examples.)

Table A-2. CLOSED FORM OF $\sum\limits_{m=1}^{\infty} f(m)$

$f(m)$	m	Closed Form	
$1/m^2$	all	$\pi^2/6$	
$1/m^2$	odd	$\pi^2/8$	
$(-1)^{m+1}/m^2$	all	$\pi^2/12$	
$1/m^4$	all	$\pi^4/90$	
$1/m^4$	odd	$\pi^4/96$	
$1/(m^2 + a^2)$	all	$\frac{1}{2}[(\pi/a) \coth \pi a - 1/a^2]$	$a \neq 0$
$\sin mx/m$	all	$\pi - x/2$	$0 \leq x \leq 2\pi$
$\sin mx/m$	odd	$\pi/4$	$0 \leq x \leq \pi$
$\cos mx/m^2$	even	$x^2/4 - \pi x/4 + \pi^2/24$	$0 \leq x < \pi$
$\cos mx/m^2$	odd	$-\pi x/4 + \pi^2/8$	$0 \leq x < \pi$
$(-1)^{m+1} \sin mx/m^3$	all	$(x/12)(\pi^2 - x^2)$	$-\pi \leq x \leq \pi$
$\dfrac{2m \sin 2mx}{(2m - 1)(2m + 1)}$	all	$(\pi/4) \cos x$	$0 \leq x < \pi$
$\cos mx/4$	odd	$\pi^4/96 - (\pi^2/16)x^2 + (\pi/24)x^3$	$0 \leq x < \pi$
$\sin mx/m^5$	odd	$(\pi^4/96)x - (\pi^2/48)x^3 + (\pi/96)x^4$	$0 \leq x \leq \pi$

REFERENCES

1. LANGHAAR, H. L., *Energy Methods in Applied Mechanics*, John Wiley & Sons, Inc., New York, 1962.

2. ARGYRIS, J. H., and KELSEY, S., *Energy Theorems and Structural Analysis*, Butterworth Scientific Publications, London, 1960.

3. DYM, C. L., and SHAMES, I. H., *Solid Mechanics: A Variational Approach*, McGraw-Hill Book Co., New York, 1973.

4. FUNG, Y. C., *Foundations of Solid Mechanics*, Prentice-Hall, Inc., Englewood Cliffs, N. J., 1965.

5. HOFF, N. J., *The Analysis of Structures*, John Wiley & Sons, Inc., New York, 1956.

6. PRZEMIENIECKI, J. S., *Theory of Matrix Structural Analysis*, McGraw-Hill Book Co., New York, 1968.

7. ODEN, J. T., *Mechanics of Elastic Structures*, McGraw-Hill Book Co., New York, 1967.

8. PESTEL, E. C., and LECKIE, F. A., *Matrix Methods in Elastomechanics*, McGraw-Hill Book Co., New York, 1963.

9. RAYLEIGH, J. W. S., *Theory of Sound*, Dover Publications, Inc., New York, 1945.

10. TIMOSHENKO, S. P., and GERE, J. M., *Theory of Elastic Stability*, McGraw-Hill Book Co., New York, 1961.

11. RITZ, W., "Ueber eine neue Methode zur Lösung gewisser Variationsprobleme der mathematischen Physic," *J. Reine Angew. Math.* Vol. 135, pp. 1–61, 1909.

12. COURANT, R., and HILBERT, D., *Methods of Mathematical Physics*, Vol. 1, Interscience Publishers, New York, 1953.

13. GELFAND, I. M., and FOMIN, S. V., *Calculus of Variations*, (translated from the Russian, and edited by R. A. Silverman) Prentice-Hall, Inc. Englewood Cliffs, N. J., 1963.

14. KANTOROVICH, L. V., and KRYLOV, V. I., *Approximate Methods of Higher Analysis*, 4th ed. (translated from the Russian by C.D. Benster), Interscience Publishers, New York, 1958.

15. BERG, P. W., "Calculus of Variations," Ch. 16, *Handbook of Engineering Mechanics* (edited by W. Flügge), McGraw-Hill Book Co., New York, 1962.

16. PHILLIPS, E. G., *Functions of a Complex Variable with Applications*, Oliver and Boyd, London, 1954.

17. CARSLAW, H. S., *Introduction to the Theory of Fourier's Series and Integrals*, Dover Publications, Inc., New York, 1930.

18. FRANKLIN, P., "Basic Mathematical Formulas," in *Fundamental Formulas of Physics* (edited by D. H. Menzel), Dover Publications, Inc., New York, 1960.

INDEX

AUTHOR INDEX

247

SUBJECT INDEX

N

Nonconservative loads, 217, 223
Nonconservative systems, 17, 217
   Beck problem, 220
   follower force, 218, 220
   under static forces, 217

P

Postbuckling considerations:
   Koiter's theory, 151-52
Principle of:
   minimum potential energy, 234
   stationary value of the total
      potential, 232
   virtual displacements, 228
   virtual work, 228

R

Rayleigh quotient, 130
Rayleigh-Ritz method:
   for stable equilibrium
      problems, 236-40
   for unstable equilibrium
      problems, 138-43
Rayleigh-Timoshenko method, 132-38
Ritz method, 138-43, 236-40

S

Southwell Plot, 68
Special functions:
   Dirac δ-function, 52, 57
   doublet function, 54, 57
   Macauley's bracket, 52
   unit step function, 52
Stability:
   concept of, 5
   criterion for:
      dynamic (kinetic), 9
      energy, 9, 125-27

      static (bifurcation, equilibrium), 7
      Trefftz, 127
   models:
      one-degree-of-freedom, 21-30,
        34-38
      snapthrough, 34-38
      two-degree-of-freedom, 3-34
      with imperfections, 38-42
Stability analysis:
   bifurcation approach, 7
   dynamic approach, 9
   energy approach, 9
Strain energy, 225

T

Tilting of forces, 68
Timoshenko beam (*see* Beams,
      Timoshenko)
Timoshenko's method, 127
Timoshenko quotient, 130
Trefftz criterion:
   columns by the, 143-46
   low arches by the, 207-14
   for stability, 127

U

Unit-displacement theorem, 235

V

Virtual displacements:
   definition, 228
   principle of, 228-32
Virtual work:
   principle of, 228-32

W

Work principles and theorems, 225-41